T0187079

A BRIEF HISTORY OF THE LAST 13.8 BILLION YEARS

A BRIEF HISTORY OF THE LAST 13.8 BILLION YEARS

a journey through life, the universe, and everything

David Baker

Foreword by John Green

SCRIBE
Melbourne • London

Scribe Publications
2 John St, Clerkenwell, London, WC1N 2ES, United Kingdom

Published by Scribe 2022
Published by arrangement with Black Inc.

Copyright © David Baker 2022

All rights reserved. Without limiting the rights under copyright reserved above, no part of this publication may be reproduced, stored in or introduced into a retrieval system, or transmitted, in any form or by any means (electronic, mechanical, photocopying, recording or otherwise) without the prior written permission of the publishers of this book.

The moral rights of the author have been asserted.

Text design by Dennis Grauel
Typesetting by Tristan Main
Maps and graphs by Alan Laver

Printed and bound in the UK by CPI Group (UK) Ltd, Croydon CR0 4YY

Scribe is committed to the sustainable use of natural resources and the use of paper products made responsibly from those resources.

978 1 915590 02 2 (hardback edition)
978 1 761385 00 1 (ebook)

A catalogue record for this book is available from the British Library.

scribepublications.co.uk

Contents

To David Christian

Foreword

JOHN GREEN

HUMANS LIKE A GOOD STORY and, since we are as a species somewhat narcissistic, we're especially fond of the story of ourselves – how we came to be here and why. These days, we call that story history, but for too long we've had a narrow definition of history, one that dramatically distorts reality.

As a high school student, I was taught that 'recorded history' began about 5000 years ago, with the invention of writing. However, that definition leaves out almost all of the human story – at least 95 per cent of it. Of course, we cannot know the humans of 100,000 years ago as intimately as we know Genghis Khan or Cleopatra but omitting them from the story makes the human story seem much newer than it actually is. When you imagine that our story begins with the emergence of agriculture, or writing, or any particular innovation, the human story seems to look like an ascending line: lives get longer. People grow less hungry, and less poor, and more educated. Technological improvements are shared more widely, and innovations pile upon innovations to ensure that life inevitably improves.

For most of human history, that wasn't really the story. Important innovations were made, as small communities passed down knowledge from one generation to the next, but human lives have not always grown consistently healthier or more productive. As you will learn in this book, we nearly went extinct long before we figured out how to develop agriculture or steam engines or antibiotics. We have been the dominant species on this planet for the briefest flicker of our history, and until we understand that, we can't reckon meaningfully with the dramatic and sudden changes we're making to our planet and its biosphere.

Narrow views of history also too often create a false dichotomy between the 'hard' sciences – chemistry, physics, biology – and the 'soft' humanities – history and literature and anthropology. Human stories cannot be viewed in isolation – we can't understand fourteenth-century Europe without learning about the biology of *Yersina pestis* and the rats that carried it. And we can't understand how life came to be on Earth without glimpsing how time began in the first place, and how each of us is made from stars.

In *A Brief History of the Last 13.8 Billion Years*, David Baker introduces us not only to the history of our species and our planet but also to the history of our vast universe. We are not the end of that story, nor its beginning – instead, we have emerged in the middle of a tale that will continue long after we are gone. Glimpsing the breadth of the universe's history can make a person, or a species, feel very small indeed. And yet it is also a reminder of how wondrous life is, and how astonishing. As Baker writes, when we look into the night sky, we are not looking at the Universe, we are the Universe looking at itself.

Introduction

THIS BOOK FOLLOWS THE CONTINUUM of historical change of all the 'stuff' in the cosmos, from the Big Bang to the evolution of life to human history, as simple clumps of hydrogen gas are transformed into complex human societies. History allows us to live many lives instead of just one, and this particular story instils in us the combined experience of billions of years. Much confusion over human identity, our philosophies and our future could be resolved if only the average person knew the key beats of the 'story of everything' at least as well as they knew the key moments of their national histories.

Zooming out to take a bird's eye view of 13.8 billion years allows us to see beyond the chaos of human affairs to the overall shape and trajectory of history. The thread that runs through the entire grand narrative is the **rise of complexity** in the cosmos, from the first atoms to the first life to humans and the things we have made. It allows us to traverse eons without drowning in detail – because the amount of detail an answer requires depends on the nature of the question. In this book, we ask simply: where did we come from and where are we going?

Regarding the future, I am speaking in terms of the next hundred years, the next thousand, the next million, the next billion and even the next trillion, to quadrillion, to the potential end of the Universe. *The Shortest History of the World* explores all that too.

For the science-phobic, rest assured there will be no maths equations, and foreign cosmic phenomena will be boiled down to plain speech. For the history buff, humans may occupy only what a colleague called the 'thinnest chip of paint at the top of the Eiffel Tower' of 13.8 billion years, but for very real, objective reasons humans occupy a very important role in the story. As far as we know, human societies and technologies are thus far the most complex structures in the entire Universe. We are a tightly

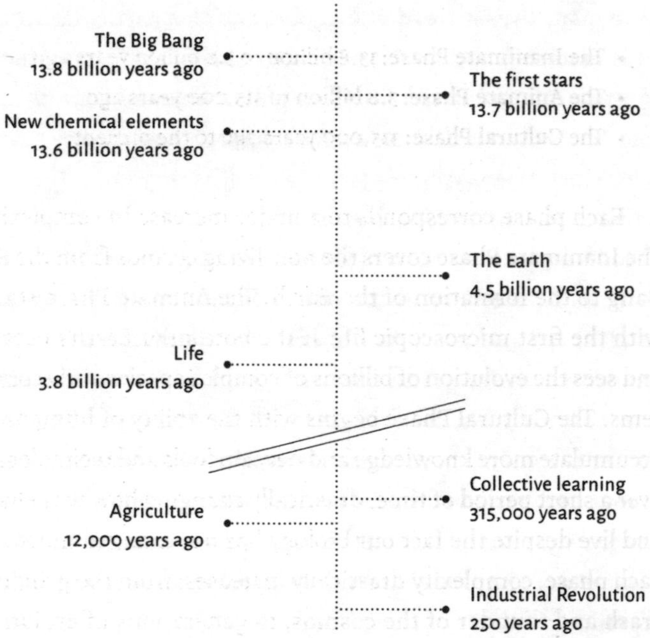
The Big Bang
13.8 billion years ago
The first stars
13.7 billion years ago
New chemical elements
13.6 billion years ago
The Earth
4.5 billion years ago
Life
3.8 billion years ago
Collective learning
315,000 years ago
Agriculture
12,000 years ago
Industrial Revolution
250 years ago

woven web of 8 billion whirring brains, each with more nodes and connections than there are stars in the Milky Way. The next rise of complexity is likely to come from us – or at least something like us that evolved elsewhere in the cosmos.

French historian Fernand Braudel once likened the political events of recent history to bubbles and swirling foam atop an ocean of deep time. Here today, gone tomorrow. To truly understand where we are and where we are going, we must look below to the deeper currents and tides. The tendency of complexity to increase in the Universe moves the entire historical ocean. This trend towards rising complexity created us and continues to change us. Astoundingly, a self-aware humanity currently has the power to control where complexity goes next.

Our past can be divided into three phases.

- The Inanimate Phase: 13.8 billion to 3.8 billion years ago
- The Animate Phase: 3.8 billion to 315,000 years ago
- The Cultural Phase: 315,000 years ago to the present

Each phase corresponds to a major increase in complexity. The Inanimate Phase covers the non-living cosmos from the Big Bang to the formation of the Earth. The Animate Phase starts with the first microscopic life at the bottom of Earth's oceans and sees the evolution of billions of complex species and ecosystems. The Cultural Phase begins with the ability of humans to accumulate more knowledge and develop tools and technologies over a short period of time, drastically changing how we behave and live despite the fact our biology has not changed much. In each phase, complexity drastically increases: from the grinding crash and thunder of the cosmos, to generations of evolution by natural selection, to cultural evolution or **collective learning**. The pace of historical change also accelerates rapidly: cosmic changes can take billions of years and evolutionary changes millions of years, whereas cultural changes are measured in millennia, centuries, years – even days.

Every shift in complexity, each major event of the past, every newly emerged form of evolution builds on what came before it.

Our story also has a fourth phase – the Unknown, in which complexity will leap forward yet again and set off an entirely new stage of cosmic evolution and historical change. Perhaps humans will give way to the accelerated creation and innovation powers of self-aware AI (artificial intelligence). Perhaps humans will upload their consciousness to computers and travel across the galaxy. Perhaps quantum physics will allow unprecedented manipulation of the building blocks and fundamental laws of the Universe. All we know for certain is that unless complexity is outright destroyed, some sort of increase of complexity is only a matter of time. And in the human realm, the changes keep coming faster and faster.

The generations of humans alive today occupy a pivotal role in a story that has been unfolding for 13.8 billion years. By understanding the long-term story from billions of years ago, we are better placed to make long-term plans billions of years into the future.

PART ONE

THE INANIMATE PHASE

13.8 billion to 3.8 billion years ago

1

THE BIG BANG

Wherein all the 'stuff' in the Universe appears • Space appears to give us somewhere to put all that 'stuff' • Time appears and makes it possible for that 'stuff' to change form (i.e. have a history) • All that 'stuff' is primordial energy and matter that transforms into the diverse range of things around us.

BANG.

13.8 billion years ago, a tiny, hot, white speck appeared. It was so small that at first it could not have been seen by the naked eye or anything except the most powerful of modern microscopes, had they existed.

This was the appearance of the space-time continuum and the extremely hot, densely packed energy within it. Nothing existed outside of it. All the ingredients for everything in the Universe were within it. They have simply changed form since then, as if the Universe were a ball of clay, shaped and reshaped into myriad forms over billions of years.

The absolute first date in all of history is 10^{-43} seconds after the Big Bang, or 1.0 with the decimal place moved to the left forty-three times:

0.0000000000000000000000000000000000000000001

A tiny sliver of a second. It is the smallest possible chunk of time that we can measure. A smaller fraction of time would be physically meaningless, because nothing in the Universe can move fast enough to show that even the slightest change has happened in a smaller amount of time. 10^{-43} seconds is the amount of time it takes for light to travel the smallest amount of distance at the quantum level. Any smaller snapshot of time (for instance, 10^{-50} seconds) looks exactly the same as 10^{-43} seconds. It is like the first frame of a film.

The Universe was smaller than an atom or even one of the particles that make up that atom. Because of the pressure of everything in the Universe being contained within that small space, it was incredibly hot. Up to 142,000,000,000,000,000,000,000,000,000,000 Kelvin or 142 novillion (so hot it is practically the same in both Celsius and Fahrenheit). The laws of physics themselves could not stay coherent. The Universe was so hot that the very laws that make it work were in a 'melted' form. It was true, unadulterated chaos. *Alice in Wonderland* and a pint of LSD.

A tiny fragment of a second later by 10^{-35} seconds after the Big Bang, the Universe had expanded to the size of a grapefruit. It would have become visible to the naked eye. It cooled below 11.3 octillion Kelvin. This was cool enough for the four fundamental forces of physics to 'harden' into their current form. Gravity, electromagnetism and the strong and weak nuclear forces became coherent. We were now a Universe governed by physical laws. If they had hardened into a slightly different balance, the Universe would have evolved completely differently.

During this time, a ripple at the quantum level made tiny pinpricks of energy clump together. Energy in the Universe was just *ever so slightly* unequally distributed. These clumps of energy would evolve into all the matter, complexity, stars, planets, animals and 'stuff' in the Universe, including you.

By 10^{-32} seconds after the Big Bang, the Universe was about a metre wide, and the heavy lifting was over. The clock was wound, its mechanisms were set in motion and it began to tick. In the first split second, our destiny was already etched into the very fabric of the cosmos. And the rest, as they say, is history.

For the next 10 seconds, the Universe grew to 10 light years wide, swirling with tiny particles which had congealed from pure energy as the Universe continued to cool to 5 billion Kelvin. They were quarks and anti-quarks, positrons and electrons. The opposites of each other. Matter and antimatter. Much of the matter bumped into the antimatter and exploded in a flash, turning back into energy. Only one-billionth of the matter could not find an antimatter partner, and it is only this tiny fraction of matter that forms all the 'stuff' in the Universe we see today. Here, within the first 10 seconds of the story, is a miracle that saved us from non-existence.

Over the next three minutes, the Universe continued to expand. It was over 1000 light years wide: a sea dominated by thick, merciless radiation. The surviving quarks were forged together by the still-intense heat into protons and neutrons. These protons and neutrons were in turn forged into the core of hydrogen and helium atoms (the nucleus). Hydrogen and helium were the simplest and first of all the elements to exist. Hydrogen requires only a single proton as its nucleus. Helium requires more ingredients and was therefore in the minority. The Universe cooled below 100 million Kelvin – too quickly for many of the other elements to be created (only trace amounts of lithium and beryllium). Heavier elements would have to wait for the creation of stars many millions of years later.

The Universe continued to expand and cool for thousands of years, longer than *Homo sapiens* has existed. By 380,000 years after the Big Bang, the Universe was over 10 million light years wide and it had cooled to 3000 Kelvin – twice as hot as lava and enough to melt gold or cause a diamond to drip like an ice cube on a summer's day. The heat was still enough to obliterate most complexity, but it was cool enough for hydrogen and helium nuclei to capture electrons and become fully-fledged atoms. The Universe began to fill with clouds of gas.

The Universe had also become less dense, allowing photons of light to travel freely through the thick soup of radiation and particles for the first time. There was a blinding flash of light, as these photons headed in every conceivable direction. This

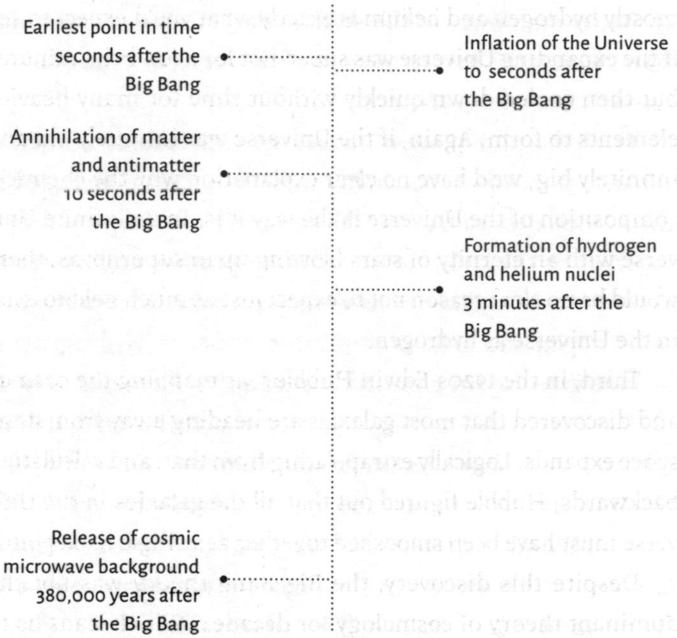
Earliest point in time
seconds after the
Big Bang
Inflation of the Universe
to seconds after
the Big Bang
Annihilation of matter
and antimatter
10 seconds after
the Big Bang
Formation of hydrogen
and helium nuclei
3 minutes after the
Big Bang
Release of cosmic
microwave background
380,000 years after
the Big Bang

flash of light is known as the cosmic microwave background (CMB) and can be detected in every direction in the Universe today. In fact, if you set your radio or TV to collect only static, about 1 per cent of that static will be from the CMB. It is the first baby picture of the Universe and the first visible artefact of our deep past.

HOW DO WE KNOW THE BIG BANG HAPPENED?

We know the Big Bang happened for several reasons. For starters, we can't find anything in the Universe – either on Earth or through a telescope – that is confirmed to be older than 13.8 billion years old, which is the current estimated age of the Universe. If the Universe was infinite and eternal, we'd be tripping over stuff that was 105 billion or 802 trillion years old.

Second, the fact that normal matter in our Universe is mostly hydrogen and helium is exactly what you'd expect to see if the expanding Universe was super-hot for a few brief minutes but then cooled down quickly without time for many heavier elements to form. Again, if the Universe was infinitely old and infinitely big, we'd have no clear explanation why the chemical composition of the Universe is the way it is. In an infinite Universe with an eternity of stars blowing up in supernovas, there would be no clear reason not to expect just as much gold to exist in the Universe as hydrogen.

Third, in the 1920s Edwin Hubble was mapping the cosmos and discovered that most galaxies are heading away from us as space expands. Logically extrapolating from that, and calculating backwards, Hubble figured out that all the galaxies in the Universe must have been smooshed together at a single fixed point.

Despite this discovery, the Big Bang theory was not the dominant theory of cosmology for decades. Which leads us to the fourth and most crucial piece of evidence: the cosmic microwave background that emerged 380,000 years after the Big Bang. If the Big Bang theory were true, then after a few thousand years of the Universe expanding, the crush of matter and plasma and radiation would be spread out enough for light to be able to travel freely, and there would be a brilliant flash across the cosmos. In the 1940s, physicists predicted we should be able to see the remnants of this flash everywhere in the sky. This is precisely what was discovered in 1964 by two radio engineers, Arno Penzias and Robert Wilson, who weren't even looking for it. They were trying to eliminate all the static on a highly sensitive radio antenna, but they couldn't get rid of a small hiss, and after many calibrations, and having shot the pigeons that shat on the antenna, a physicist from Princeton told them what they had found. From that point on, the Big Bang became the dominant explanation for the start of the Universe and all the work that has happened since has only confirmed or clarified the general framework of this theory.

WHAT DOES THE UNIVERSE LOOK LIKE?

In the first split second after the Big Bang, the Universe inflated from the size of a quantum particle to that of a grapefruit. Within a second, it was larger than our solar system. Four years later, it was bigger than the Milky Way.

The Universe, as we know it, is currently 93 billion light years across. Which means there are stars and galaxies that were born billions of years ago that are so far away their light hasn't had a chance to reach us, since only 13.8 billion years have elapsed since the start of the Universe. The stuff we can see looking out from Earth is called the Observable Universe, but there are a lot of things beyond that horizon which we cannot see.

Moreover, because light takes time to travel from a distant object, the further away we look, the further we are looking into the past. For instance, a neighbouring galaxy, Andromeda, is 2 million light years away. So when you look at it through a telescope, you are seeing it as it existed roughly when *Homo erectus* started roaming the Earth and sabre-toothed tigers were still a cause for concern.

The Observable Universe can be seen looking in any direction from Earth; in that sense, the Observable Universe is a sphere. However, that is not the shape of the *entire* Universe. Physicists have determined that the Universe has 'zero curvature', which means it does not bend back on itself at some point. It stretches on and on, like a table-top in any direction, constantly expanding for all eternity. The Observable Universe is just one patch on

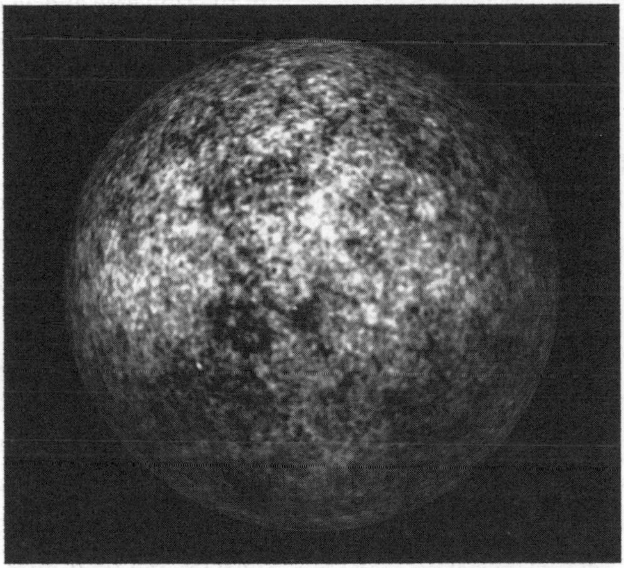

Cosmic microwave background (CMB)

it: like the ring left by a coffee cup on a table. And Earth is just one tiny fibre of wood lodged somewhere inside that coffee ring.

The colour of the Universe is beige, assuming we were looking at the entire Universe from a great distance with human eyes. If you were to look at the mixture of light from all the stars in the Observable Universe blended together, as if you were zoomed out and looking at the whole thing at once, the colour of our cosmic bubble would be beige. Cosmologists have tried to jazz it up by calling the tone 'cosmic latte', but it's really just beige. Personally, I like the fact the Universe is beige; it renders the cosmos slightly less intimidating.

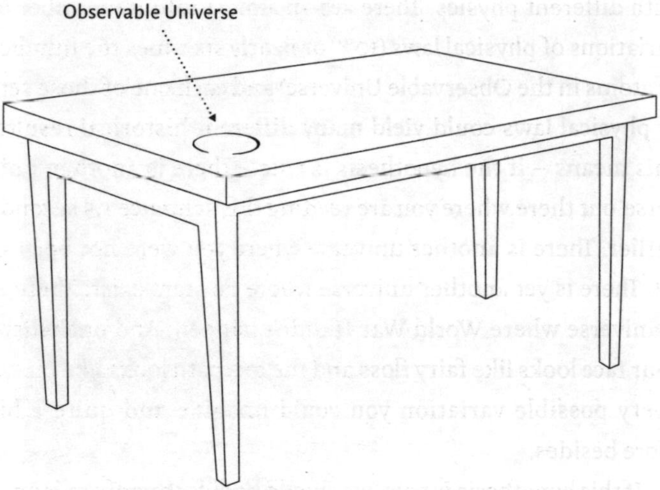
Observable Universe

The Observable Universe

WHAT IS THE MULTIVERSE?

Permit me to get a little weird for a moment. One inevitable consequence of the Big Bang model (the most accepted model currently) is a phenomenon called 'eternal inflation', which means that while our coffee ring of the Observable Universe has popped out of inflation and is expanding more slowly than in the first split second, other parts of the table-top may still be expanding at that speed. And there will be other coffee rings (that is, other so-called universes) with physical laws and variations in historical events entirely different to our own. And this process would stretch on forever. This collection of diverse 'universes', each roughly the size of our 'Observable Universe', is known as the Multiverse.

But the term Multiverse is a misnomer: it's all the same Universe, just different patches or coffee stains on the table, with different physics. There are an almost infinite number of variations of physical laws (10^{500} or nearly six times the number of atoms in the Observable Universe) and each one of those sets of physical laws could yield many different historical results. This means – if the hypothesis is true – there is another 'universe' out there where you are reading this sentence 1.5 seconds earlier. There is another universe where you were not born at all. There is yet another universe where no stars exist. There is a universe where World War II didn't happen. And one where your face looks like fairy floss and the footpath looks like pizza. Every possible variation you could imagine and quite a bit more besides.

If this hypothesis is true, we should be able to confirm it once the light from the nearest other 'universes' that have appeared (if they have such a thing as light) finally reaches us ...

... in about 3 trillion years.

HOW CAN WE UNDERSTAND THE BIG BANG?

If trying to understand how our Universe began gives you an existential headache, it's not your fault. Humans evolved within an older Universe with fixed rules, as did our brains and perceptions, so it is not so easy for our brains to grasp an event that precedes the establishment of the physics we have intuitively come to know. We are evolved to instinctually understand the world enough in order for our species to survive: what goes up must come down, cause and effect, chickens come from eggs and eggs from chickens. The rest takes a bit more time and reflection.

Imagine a speck. A tiny speck. This is the Big Bang singularity 13.8 billion years ago at 10^{-43} seconds. All energy and matter were contained within that speck. All the ingredients for the rest of our story. But whatever you do, don't imagine that there is space outside the speck. Space is a property of our Universe and exists entirely inside it. As the Universe expands, more space is created. Don't even imagine pitch blackness outside the speck, like we see at night between the stars. That is space. At the moment of the Big Bang there was nothing but the speck.

In fact, get a piece of paper and a pen and draw a tiny dot in the middle of the paper. Then take a pair of scissors and cut off all the extra paper outside the dot. That is the early Universe. The primordial atom containing all of time, space and energy, which has grown into the tabletop that is still expanding today.

WHAT HAPPENED BEFORE THE BIG BANG?

Time didn't exist before the Big Bang; thus, there was no 'before' the Big Bang. It would be like claiming you introduced your mother and father to each other: nonsensical.

Space also did not exist before the Big Bang. 'Before' the Big Bang, there was no space for anything to happen and there wasn't any time for it to happen either. After the Big Bang, the Universe expanded from the microscopic to its current size of 93 billion light years across (and growing). Space is a post–Big Bang phenomenon. Time is post–Big Bang too. 'Before' the Big Bang, if there is no space for anything to move, there is no space for anything to change. And if there is no change, there are no events, and no history. Nothing that could be measured by time in any meaningful way.

So 'before' the Big Bang there was no space, no change and no 'stuff' that moved or was transformed. Nada, zip, zilch. If anything existed prior to the Big Bang, it would have behaved in a way completely foreign to humans, and to the fundamental laws of the Universe itself as we now know it. It would not have behaved in a sequence of cause and effect – of past, present and future.

Hence, our history begins with the Big Bang.

HOW DO YOU GET SOMETHING FROM NOTHING?

There's a deeply ingrained bit of human logic that says: if you create something, its building blocks have to come from somewhere else. This is what the First Law of Thermodynamics boils down to: matter and energy are neither created nor destroyed, they simply change form. Yet the Universe seems to have appeared out of nowhere.

But at the moment of the Big Bang, the Universe was so insanely hot (142 novillion Kelvin), its physical laws did not yet exist. This includes the First Law and the general notion that something has to come from somewhere.

Furthermore, the Big Bang at 10^{-43} seconds was so small it was at the quantum scale. Things operate differently in the quantum realm. Little ripples of energy known as virtual particles appear and disappear at that scale all the time. They are currently doing so in between the atoms that form your skin. Popping in and out of existence as if from nowhere. And this is established physics *within* our Universe, so 'something from nothing' is not really so unthinkable a proposition for the start of the cosmos. Perhaps our Universe emerged in a similar way to virtual particles.

There's also the consideration that before the existence of time, you don't have the traditional sequence of cause-and-effect with which humans have evolved and expect to see. There is no physical law obliging the Universe to have emerged from something else.

And further still, we humans don't really know what nothing is – beyond the meanings we ourselves have invented. As a shorthand expression, 'nothing' means the absence of something specific. The concept of nothing works in the context of I have 'nothing' in my pint glass and 'nothing' in my wallet to buy another beer. Within true blue physics, however, it is impossible for 'absolutely nothing' to exist anywhere in the Universe – even in the deepest regions of space. Everywhere the Universe either has 'stuff' like stars and planets and gas or at the very least the weak hum of radiation. Your wallet may not have money in it, but it has air, a debit card, some old ticket stubs, dust, and perhaps even a dead fly. Scientists can't even create artificial spaces where there is truly nothing. It is physically impossible to create what is called a 'zero energy vacuum' or a void that doesn't even have radiation. So where does 'nothing' actually exist? We appear to have invented 'nothing' out of thin air.

'Nothing' being physically impossible in our Universe, we are making a huge assumption and leap of logic that 'nothing' (a concept humans invented and cannot replicate) existed 'before' the Big Bang. In fact, the grammar of that statement is all wrong. We have no reason to expect that 'nothing' as a concept truly exists somewhere outside the Universe and that it preceded the Big Bang when time didn't even exist yet. By saying 'something from nothing' we are making huge assumptions that we are not scientifically or logically entitled to make.

We have to unlearn some of our most basic concepts to understand the workings of a primeval Universe without the same rules it has now. The primate brain runs up against concepts we didn't need to understand in order to survive and evolve. Our brains aren't wired that way. It is like trying to send your friend a text using your toaster.

SEEKING ANSWERS AT THE BEGINNING

If you find yourself hit by a wave of existential disquiet and dissatisfaction about the mysteries of the Big Bang, consider the following:

1. We did not even confirm the Big Bang happened until sixty years ago. Imagine how many answers we will discover about the start of the Universe after another 100 or even 1000 years of scientific endeavour.

2. If the answers to this puzzle are foreign to our primate brains and foreign to the fundamental physics of this Universe, then the answers (when we discover them) might sound like gibberish to us. They may not satiate the emotional and philosophical void and search for meaning we have assumed they would fill.

3. We may be looking for satisfaction in the wrong place by looking to the beginning of the story. Perhaps if

> we seek to add meaning to our lives, we must look at the present or perhaps even towards how we would like the story to end. In our own lives, we at least enjoy some measure of control over our own destiny. And if humanity continues to exist, our science and technologies continue to develop and our overall complexity continues to increase, who knows what profound superhuman impact we will have on the story in a thousand, a million or a billion years?

Philosophical satisfaction and existential meaning are frequently derived not from obsessing over our childhood traumas or what happened in the world before we entered it, but by making good and honourable use of the time given to us. If the earliest moments of the Universe prove anything, it is that seemingly tiny changes can be writ large into the fabric of the cosmos.

2

STARS, GALAXIES AND COMPLEXITY

Wherein the first hydrogen and helium atoms get sucked together into clouds • Those clouds become so tightly packed that the atoms fuse together • Fusion creates huge nuclear explosions that birth the first stars • Stars fuse hydrogen and helium into carbon, nitrogen, oxygen and other elements, up to the twenty-sixth element of iron • The stars blow up in supernovas, creating heavier elements such as gold, silver and uranium • All ninety-two naturally occurring elements are created by these terrifying H-bombs in the sky.

AS WE HAVE SEEN, 10^{-43} SECONDS AFTER THE BIG BANG, the Universe rapidly expanded from the size of a quantum particle to the size of a grapefruit. For perspective, if the same rate of expansion had continued to happen, that grapefruit would have grown to the size of the current Universe in a fraction of a second rather than taking 13.8 billion years to do so. During that split second, tiny inequalities in energy appeared. Just a little bit more energy in dots sprinkled all over the cosmic grapefruit, differing from the almost equal distribution of energy found elsewhere in the Universe. It is those tiny dots with just a little bit more energy that spawned stars, galaxies, planets and all the complexity that forms our history. Without these inequalities, complexity in the Universe would have been born 'dead'. And this short history would be a great deal shorter.

As energy within those dots congealed into subatomic particles, the first matter emerged and the Universe continued to expand and cool.

Filled with a cloud of hydrogen and helium gas, the Universe continued to cool as it expanded to just a little above absolute zero, where it remains today. The vast majority of space from this point forward remained simple and cold, and there was no heat to forge further complexity beyond hydrogen and helium. Most of space was filled only with weak radiation. It was only in

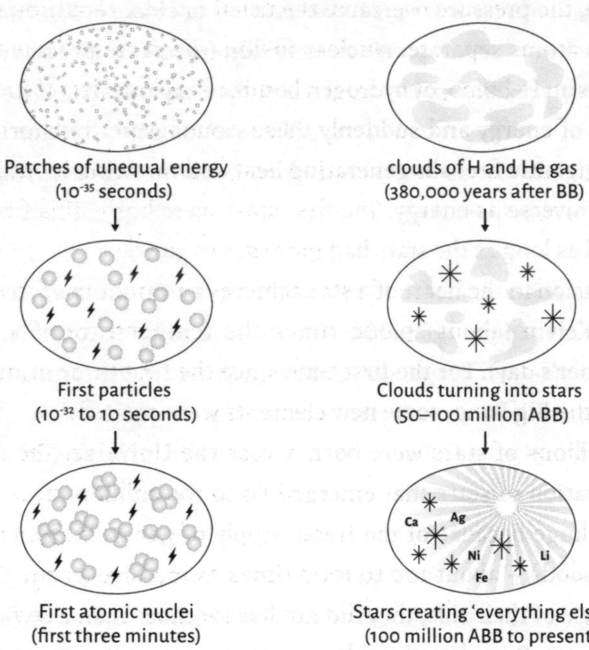
Patches of unequal energy
(10^{-35} seconds)
First particles
(10^{-32} to 10 seconds)
First atomic nuclei
(first three minutes)
clouds of H and He gas
(380,000 years after BB)
Clouds turning into stars
(50–100 million ABB)
Ca
Ag
Ni
Li
Fe
Stars creating 'everything else'
(100 million ABB to present)

tiny pockets where inequalities in matter and energy reigned that things began to heat up.

THE FIRE-SOAKED ORIGINS OF STARS

For millions of years, enormous clouds of hydrogen and helium floated through the ever-expanding cosmos. There was not much else among the gloom, and the Universe seemed pretty homogenous. Dull, dead and without much change or history.

Over the course of 50 to 100 million years after the Big Bang (or, very roughly, the amount of time that separates you from *Tyrannosaurus rex*), gravity sucked hydrogen and helium gas together into increasingly dense clouds. Eventually the pressure at the core of these clouds became so intense that hydrogen atoms were smashed together and their nuclei fused. That is to say, the pressure overcame the usual nuclear repulsion that keeps atoms separate. Nuclear fusion (the same process that causes an H-bomb, or hydrogen bomb, to explode) let off a huge burst of energy and suddenly these clouds were transformed into gigantic fireballs generating heat and throwing it out into the Universe as energy. The first stars were born. This fusion lasted as long as the stars had more gas to guzzle.

Fusion in the heart of a star achieves a minimum of 10 million Kelvin (about 25,000 times the temperature of a hot summer's day). For the first time since the first three minutes after the Big Bang, some new elements were created.

Billions of stars were born across the Universe. The first generation of stars that emerged 50 to 100 million years ago were huge because of the fresh supply of gas in their neighbourhoods – about 100 to 1000 times as massive as our Sun. Because of their size, they did not live for more than a few million years. But when they blew up, their materials were sucked back together into second-generation stars, which were smaller but could live for longer – into the billions of years.

Gravity began to attract stars to each other, and they formed clusters that were 30 to 300 light years across. These clusters merged into even bigger ones. From 13.7 to 10 billion years ago, these mergers continued in our region of space to create the Milky Way, which is roughly 100,000 light years across. Our galaxy contains about 200 billion stars. And the same process of galactic mergers occurred all over the Universe, forming the estimated 400 billion galaxies that exist in the Observable Universe.

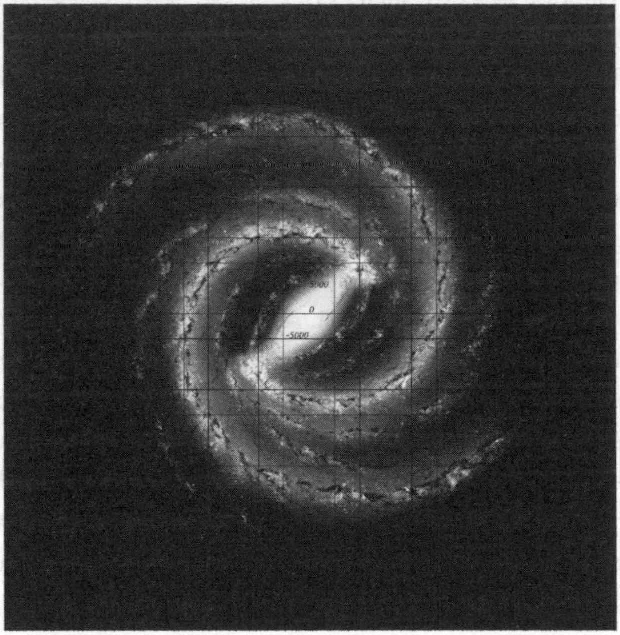
5000
0
-5000

The Milky Way

THE CREATION OF GALAXIES

From 13 billion to 10 billion years ago, different kinds of galaxies came into being. Spiral galaxies (like our own Milky Way) form 60 per cent of the estimated 400 billion galaxies in the Observable Universe. They do the majority of star formation. But their bulbous cores contain such a concentration of stars that they are hostile to the formation of life, with supernovas ripping through the sector far too frequently. It is only at the arms of spiral galaxies, where our solar system has drifted, that star systems are distant enough from each other to form life.

Lenticular galaxies (such as the Sombrero Galaxy) have the same bulge but no arms. They make up roughly 15 per cent of galaxies in the Universe. They have very little star formation.

Elliptical galaxies (such as Hercules A) have no bulge at the centre and stars are more evenly distributed. They are 'dying' galaxies and very little new star formation happens within them. They constitute 5 per cent of all galaxies in the Universe.

Irregular galaxies are the hodgepodge of malformed galaxies which are not easily categorised. They make up 20 per cent of all galaxies. Most of them are quite small and are usually misshapen by the gravitational pull of another galaxy while they are forming. Some currently have no explanation.

As for the number of galaxies in the Observable Universe, 400 billion is the established estimate. However, recent studies suggest that the number could be between 1 and 10 trillion. This – considerably higher – number increases the odds that complex life might be evolving somewhere else out there. Each galaxy can contain millions, billions or even trillions of stars. That is quite a few rolls of the evolutionary dice.

THE LIFESPAN OF STARS

The size of a star determines how long it lives, because of how quickly it burns its fuel. Stars that are over eight times the size of the Sun go supernova. Stars that are smaller than that die without exploding and creating heavier elements. The largest stars burn for only a few million years, slightly smaller stars might burn for a few hundred million years, even smaller stars might burn for a few billion years, and the smallest, slowest-burning stars can burn for a potential 100 billion to several trillion years.

The first generation of stars to form after the Big Bang were huge and blew up billions of years ago. The second generation of stars, created from the exploded remnants, contained heavier elements that were created in the bellies of the first generation of stars. Most of the second-generation stars have also died in the past 13 billion years, but many of them are still detectable in the Universe and within the Milky Way galaxy.

The third generation of stars is only a few billion years old. They possess a vast diversity of heavy elements created in the previous two generations. The third generation is also likely to have more planets orbiting them because of the abundance of elements that would have formed a ring of dust around them and eventually turned into planets. Thus, third-generation stars (like our Sun) are the best bet for further complexity.

COSMIC FLORA AND FAUNA

Our Sun is a Yellow Dwarf, a kind of star that lasts 4 to 15 billion years and represents 10 per cent of the stars in the Universe. Slightly smaller stars, called Orange Dwarfs, live 15 to 30 billion years and make up another 10 per cent. Red Dwarfs are the smallest stars (about 5 to 50 per cent the mass of the Sun) and make up 70 per cent of all the stars in the Universe. Red Dwarfs can last hundreds of billions of years – or even trillions, depending on how small and slow-burning they are. None of these stars explode in a supernova when they die, but slowly burn down and flicker out.

When a star like our Sun burns up all its hydrogen and helium fuel, it starts burning through progressively heavier elements in its core. As a result of this process, Yellow Dwarfs bloat like a dead cow in a wet field and become Red Giants. After another billion years or so, they shrink back and become White Dwarfs – the skull and bones of stars like ours which have stopped fusing atoms in their cores. These dwarfs last for another few million years before finally flickering out completely. Red Giants and White Dwarfs make up approximately 5 per cent of stars in the Universe.

The remaining 5 per cent of stars are decidedly rarer but much more essential to complexity. These are the stars that blow up in supernovas. Supergiant stars only burn for between a few million years and a few hundred million years (depending on their size). These Supergiants are capable of fusing all the atoms in the periodic table up to iron – the twenty-sixth element. Thereafter, no star's core burns hot enough to fuse anything more. Once the Supergiants run out of fuel, their massive structures collapse in on themselves, letting off a huge explosion – a supernova. The supernova itself burns so hot that during the process even heavier elements are forged, such as gold, silver and uranium. Supernovas are responsible for producing the ninety-two naturally occurring elements in the Universe. The fact that some elements (such as gold) only exist because of the supernovas of less than 5 per cent of stars is why these elements are so rare.

When stars blow up in supernovas, they leave behind neutron stars as their dead remnants. Neutron stars are extremely dense and heavy and don't burn very brightly. If two neutron stars smash into each other, they can create even more of the heavy elements. They are also tiny, being only a few dozen kilometres across. All that mass in such a small space makes them very vulnerable to turning into black holes.

A black hole is essentially a pile of matter with such a high mass that the gravity sucks it in on itself. Their gravity begins to suck in matter around them, distorting the space in their neighbourhoods. While black holes may just be sloppy piles of matter, there are some hypotheses that black holes warp space and time around them to such an extent they may have bizarre properties. For instance, they may break down the laws of physics, make the passage of time incoherent or perhaps even link to other dimensions or other universes.

STARS WITH CHEMISTRY

The periodic table currently holds 118 elements. While ninety-two elements are found naturally around the Universe, any of the higher elements that form in nature would almost immediately degenerate into lower forms. Higher elements have been created in human laboratories, the most recent one being #118, Oganesson, created by a Russian–American team in 2002.

Complexity rose within stars as they went through their life-cycles. Then they died and flung those elements out into the Universe again. They would form the building blocks for further complexity. An almost innumerable number of combinations of chemicals. To date, there are an estimated 60 to 100 million chemicals out there.

A chemical is built upon a combination of elements strung into a higher structure: a molecule. This can create a structure such as H_2O (two hydrogen atoms and one oxygen atom) to make water, or a structure such as SiO_2 (one silicon and two oxygen) to make quartz, the most common mineral on Earth, or it can create a manmade structure such as C_2H_4 (two carbon, four hydrogen) to make polyethylene, the world's most common plastic.

Then there are more complex chemicals, such as organic proteins, which are immense tangles of thousands of atoms, like the protein dubbed 'Titin', the chemical formula of which is $C_{169723}H_{270464}N_{45688}O_{52243}S_{912}$ and which gives your muscles their elasticity. The technical name of this chemical is roughly 190,000 letters long and takes somewhere between three and four hours to fully read out loud. Such is the immense scope for complexity once elements start forming into molecules! Same goes for the chemical formulas for the bases of DNA (adenine, guanine, cytosine and thymine), which encode genetic traits and allow organic material to self-replicate, evolve and become alive.

Once the ninety-two naturally occurring elements in the cosmos emerged and started combining into different chemicals, the Universe had all the ingredients it needed to create the complexity we see around us today.

But what is complexity?

THE UNIFYING PATTERN OF ALL HISTORY

The unifying pattern of all history over 13.8 billion years is increasing complexity. It is the process that created us and is, in turn, the process through which we create. After the Big Bang, the first particles of matter appeared and slowly transformed into stars. These stars would create all the chemicals that compose the Earth (including life). The same increase of complexity defines human history – from foraging, to ancient agriculture, to modernity. It is very rare in the chaos of history to find a thread that stretches across all events from beginning to end. Increasing complexity is the only such trend that has yet been identified.

A complex thing is composed of matter, intricately woven like a tapestry. It is sustained in its shape by flows of energy that 'feed' it. For instance, stars require more gas to burn. Humans require food. Mobile phones require batteries. It is all the same principle – we need flows of energy to keep from dying. That is a general rule of all complexity throughout the Universe.

Matter and energy were born within the white-hot speck of the Big Bang 13.8 billion years ago. All the ingredients for all the stuff we see around us was there at the beginning. The history of the entire Universe boils down to a history of their perpetual transformation into new and brilliant forms.

No new matter and energy were added to the Universe after the Big Bang. This is the First Law of Thermodynamics acting in full force: nothing new is created, nor is anything old totally destroyed. That means the atoms that make up your body existed in some form at the beginning of the Universe and have continued to exist and evolve across the cosmos over 13.8 billion years. You are, after a fashion, 13.8 billion years old.

And after you die, those atoms will split off in different directions and continue to evolve in the Universe again. From a certain point of view, we *are* the Universe, one totality, and we are blessed – briefly – to be a self-aware part of it. As if the Universe were looking at itself in a mirror.

THE MECHANICS OF COMPLEXITY

Complexity is an ordered structure that is created and sustained by the flow of energy. A hydrogen atom is a structure composed of one proton and one electron. A water molecule is a structure of two hydrogen and one oxygen atoms. A human brain is a form of complexity, as is the toaster that a human brain invented. The human web of 8 billion people, involving trade and information exchange, is one of the most complex systems of all.

The greater the diversity of building blocks in a form of complexity and the more intricately it is constructed, the more complex it is. A star has a lot of hydrogen atoms in it, but it is not particularly complex; it is just a big disordered lump of them. Contrast that to a dog, which has a much more complex tangle of chemicals, DNA, liver cells, brain cells, blood vessels and highly complex respiratory, circulatory and nervous systems. Move a few thousand atoms of hydrogen from the core of the Sun to its surface and it keeps running as if nothing happened. Replace a dog's brain cells with its liver cells and the dog is not going to be chasing birds anymore.

In order for any form of complexity to be created, some energy needs to be used. Like welding a car engine together in a factory. In order for that complexity to be sustained, you need more energy flow. Like eating food in order to stop from starving and dying. And in order for something to increase in complexity, it needs more energy flows altogether. If those energy flows cease, the structure decays and the thing gradually dies. A car sputters out and stops, a plant withers away and dies, a civilisation collapses into abandoned ruins. This is also why complexity can be measured in the density of energy that is flowing through it.

The more structurally intricate a form of complexity is, the greater the amount of free energy density it requires. The simplest and oldest complexity in the Universe, like a star, doesn't require that much energy per gram, whereas the products of billions of years of biological evolution or culture burn through a higher density of energy flows.

COMPLEX SYSTEMS	ENERGY FLOWS (ERG/G/S)
The Sun	2
A Supergiant star near to supernova	120
Algae (photosynthesising)	900
Cold-blooded reptiles	3000
Fish and amphibians	4000
Multi-celled plants (e.g. trees)	5000–10,000
Warm-blooded mammals (average)	20,000
Australopithecines (early primates)	22,000
Human foragers (Africa)	40,000
Agricultural society (average consumption)	100,000
Nineteenth-century textile machine	100,000
Nineteenth-century society (average)	500,000
A Model-T automobile (c.1910)	1,000,000
A vacuum cleaner (present)	1,800,000
Modern society (average consumption)	2,000,000
Average airplane	10,000,000
A jet engine (F-117 Nighthawk)	50,000,000

THE BIRTH OF COMPLEXITY

A split second after the Big Bang, there were slight ripples in space-time (quantum fluctuations) that created clumps of energy unequally distributed across the cosmos. You can see these clumps recorded in the 'snapshot' of cosmic background radiation 380,000 years after the Big Bang. As a result of these clumps, energy congealed into the first particles of matter. If it weren't for this unequal distribution of energy, complexity would not exist.

In order for complexity to exist, you need energy flows to create and sustain it. In order to have energy flow, you need to have flow from where there is *more* energy to where there is *less*.

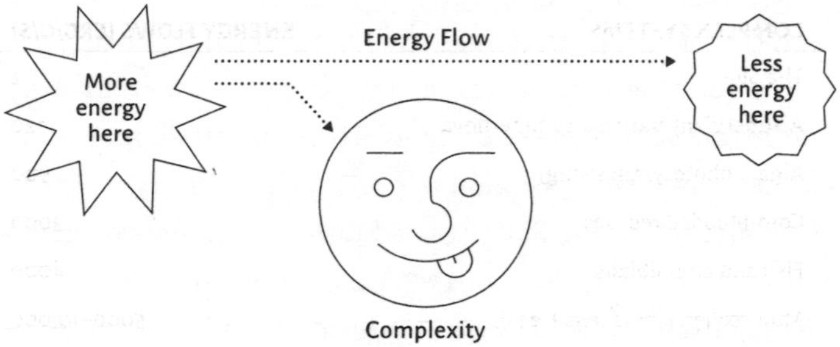
Energy Flow
More
energy
here
Less
energy
here
Complexity

If all energy were equally distributed at the start of the Universe, there would be no need for energy to move. Nothing would have changed. Nothing would have happened. There would have been no complexity, just a blank cosmos of thinly distributed radiation from start to finish. In a nutshell, there would have been no history.

Instead, the first clumps of unequally distributed matter and energy created the first stars. These stars created all the other naturally occurring elements in the periodic table. These elements came together to form molecules and planets. On one such planet, Earth, more of these molecules came together to create life. And some of that life evolved consciousness and the ability to invent stuff and continually tinker with and improve upon those inventions.

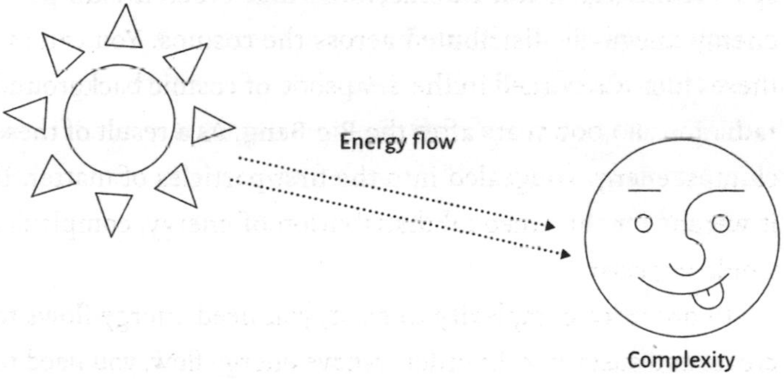
Energy flow
Complexity

All the while, from stars, to life, to technology, we required more energy flows to create, sustain and increase complexity. And so tiny pockets of the cosmos have been getting more complex over the past 13.8 billion years. That is the unifying theme of all history. The Big Bang created unequal amounts of energy across the cosmos, then for 13.8 billion years energy has been evening itself out again, and as a result of that we had energy flow and all the wondrous things that emerge from that.

THE DEATH OF COMPLEXITY

However, there is some irony to the increase of complexity in history. The reason why energy flows from stars to feed plants, which nourish animals, to give energy to brains in the human web is because of the Second Law of Thermodynamics. That law compels energy to want to even itself out – and it can only do that by flowing from where there is *more* energy to where there is *less*. In the short term, this energy flow can create complexity. But ultimately because the energy flow evens itself out, there is no more energy flow left, which kills complexity.

It is the principle which creates life and in exchange eventually takes life. Only death pays for life. This sounds like philosophy, but it is also a universal reality.

Only in tiny pockets of the Universe where there has been an unequal distribution of energy does complexity continue to rise. In the rest of the Universe, about 99.9999999999999 per cent of space is already dead, unable to generate more complexity. This is why the clumping of energy in the first split second of the Universe was so crucial to our existence.

The more complex something is, the more energy flow it requires and the faster it uses up energy flows. For example, a dog requires more energy flow per day than a tiny colony of bacteria. And a car requires so much energy that it needs to use the stockpiled energy of millions of years of organic material crushed underground and transformed into oil and petrol. Dogs poo, cars spew smoke out of their exhaust pipes, and some of that waste cannot be used again. Ever.

Eventually the Universe will run out of energy completely. After trillions upon trillions of years. So in reality, complexity is just a by-product of a longer tale in history, in which the Universe is trying to revert back to a realm of equally distributed energy. The endgame is a Universe that is nothing more than a weak orb of radiation. A quiet cosmos with no history, no change and no complexity. This is a state referred to as heat death.

The collapse of complexity is a threat throughout our story, and we will come back to the menace of heat death as we approach the end of our tale. For now, just remember that the source of our existence is also the source of our potential non-existence. The Second Law of Thermodynamics is at once the creator and the destroyer of worlds.

The only way the Second Law could be defied is by a super-civilisation many millions of years of scientific progress from now becoming so complex that it can manipulate the fundamental laws of the Universe itself.

Equal
energy
here
Equal
energy
here
Complexity

3

ORIGIN OF THE EARTH

Wherein the Sun forms and greedily sucks in 99 per cent of all matter in the solar system • The remaining 1 per cent forms a ring of dust around the Sun over a light year wide • In each orbit, the dust accretes into planets, dwarf planets, asteroids and comets • In one such orbit, the Earth forms from a series of terrifying collisions • The Earth cools and differentiation and bombardments create the first oceans • Within those oceans, long strands of organic chemicals begin to form.

OUR GALAXY, THE MILKY WAY, began from a cluster of the first giant stars about 13.5 billion years ago. Early on, it began to spin and assumed the shape of a flat disc with a bulge at the centre. As nearby galaxies were drawn into its gravitational pull, the Milky Way merged with them and grew in size. By 10 billion years ago, the last of the galactic mergers had happened. Today, our galaxy is 100,000 light years across and contains 200 to 400 billion stars.

The first generation of stars had completely died only a few million years after Milky Way had formed. The hydrogen, helium and heavy elements created in their massive supernovas were sucked together again by gravity to form entirely new stars. The second generation of stars continued to twinkle for billions of years.

Then 4.567 billion years ago, one such star, located 1 light year from where our solar system is now on a spiral arm of the Milky Way, exploded in another supernova. This explosion seeded the region with the ninety-two naturally occurring elements, from hydrogen to uranium. The burst of energy from the supernova triggered a nearby cloud of hot gas to begin the process of forming a third-generation star. Our Sun's fires flared into life for the first time. Due to the Sun's massive gravitational pull, the overwhelming majority of matter in the solar system was sucked into the Sun. The remaining 1 per cent of stuff formed a disc of tiny dust particles around the Sun, the leftovers of the process stretching on for a light year in every direction.

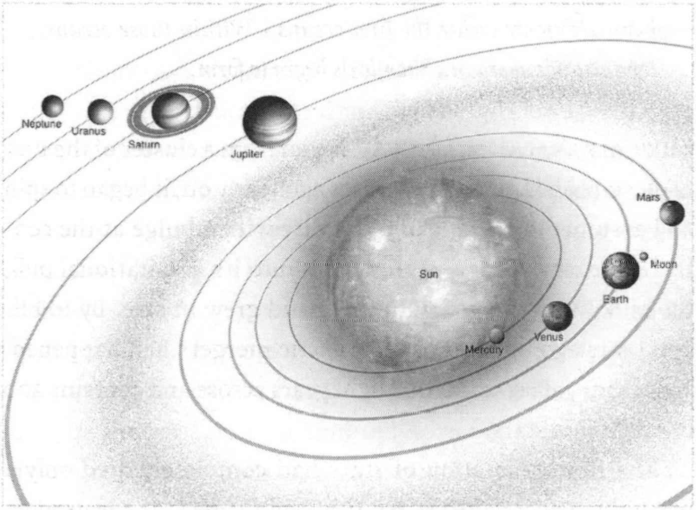
Neptune
Uranus
Saturn
Jupiter
Mars
Moon
Sun
Earth
Venus
Mercury

The Solar System

The dust of the early solar system contained all ninety-two elements, which swiftly began to form into sixty different chemicals in the vacuum of space. The Sun's first fusion ignitions blasted the majority of the hydrogen and helium gas into the outer solar system, which is why the inner planets (Mercury, Venus, Earth, Mars) are rocky and the outer planets (Jupiter, Saturn, Uranus, Neptune) are gas giants.

THE SOLAR SYSTEM

The dust around the Sun formed a flat disk and began to spin around it, much in the same way the arms of the Milky Way spin around its bulbous centre. This is the origin of our orbit around the Sun. As the dust began to spin, they formed orbital 'tracks' where everything in each orbit began to gently stick together from static electricity. In every orbital track where there is now a planet, the dust would rapidly become the size of rocks, and then the size of boulders, and then the size of mountains.

Within 15,000 years, the solar system was full of millions of objects that were over 10 kilometres (6.2 miles) across. Then the collisions became decidedly less gentle. The smashing together of these objects generated heat, which pasted the two colliding objects together. After about 10 million years, the solar system had about thirty or so proto-planets, each roughly the size of the Moon or Mars. The exception was the Asteroid Belt, where the gravitational pull of nearby Jupiter prevented numerous asteroids from colliding with each other and accreting, rendering the Belt a 'failed planet'. A few million years later, these proto-planets also collided in terrifying crashes, forming just eight planets in each orbital track:

1. Mercury is 3 light minutes from the Sun and 5 per cent the size of Earth. It suffers from extreme temperatures: –170° Celsius (–274° Fahrenheit) at night to 427°C (800°F) during the day.

2. Venus is 6 light minutes from the Sun, and the planet is a very similar size to Earth. It could have spawned life were it not for the horrific thick atmosphere of carbon dioxide, which traps a lot of heat from the Sun and sees surface temperatures hot enough to melt lead.

3. Earth is 8 light minutes from the Sun. In terms of distance, it is within the habitable zone from the Sun. Naturally, we know that conditions on Earth were right for life. We shall return to Earth shortly.

4. Mars is 12.5 light minutes from the Sun and is 10 per cent the size of Earth, which means it cannot capture much of an atmosphere – roughly 1 per cent as thick as Earth's. This means that Mars cannot maintain water in liquid form. Most water on Mars is locked away as ice, which makes life less likely.

5. Beyond the Asteroid Belt, Jupiter is 43 light minutes from the Sun, made of 99 per cent hydrogen and helium. It is eleven times the diameter of Earth and roughly 320 times Earth's mass. Jupiter has violent weather patterns which would obliterate anything resembling life. Underneath all that thick cloud, the surface may be composed of a lot of solid hydrogen; that is, hydrogen gas that is so compressed it takes on a solid

appearance. The moons of Jupiter may have fostered life, as – theoretically – the moon Europa may have, but whether or not they have remains a mystery.

6. Saturn is 78 light minutes from the Sun, nine times the size of Earth and has ninety-five times its mass. Like Jupiter, it also would not make a very good prospect for life. However, Saturn has managed to capture sixty-two moons and a ring of ice and rock, which is its signature (Jupiter also has a ring but it is much smaller). The most likely candidate for life is Titan, but life would have to have evolved very differently there than life on Earth. The temperatures are so cold that water is always frozen solid and methane gas takes the form of liquid. So if life evolved in methane oceans, it would have to have a completely different form of respiration than the life that evolved in Earth's oceans.

7. Uranus is 2.5 light hours from the Sun, four times the size of Earth and is the coldest planet in the solar system. Wind speeds are horrific, and in many ways the atmosphere and immense pressure resemble the other gas giants, making higher complexity extremely unlikely to have arisen there.

8. Neptune is 4 light hours from the Sun, the furthest planet in the solar system. It is so distant that it takes 165 Earth years to make one orbit of the Sun. Much like Uranus, Neptune is extremely cold, with a hydrogen and helium atmosphere and a core made up mostly of ice and rock.

9. Pluto used to be considered a planet since its discovery in 1930 as the furthest planet-like object we could see with telescopes at the time, approximately 5.5 light hours from the Sun. But Pluto has not cleared all the other objects from its orbit, like the eight planets have. And decades later, other dwarf planets were discovered out there, some of them bigger than Pluto, such as the dwarf planet Eris. So, sadly, Pluto was stripped of its planetary status in 2005. The fact that it shares its name with a cartoon dog has not aided this transition in status.

10. The Kuiper Belt begins 5 light hours from the Sun and stretches in a ring of planetary shrapnel as far as 7 light hours. It contains dwarf planets such as Pluto, Eris, Charon, Albion, Haumea and Makemake. The Kuiper Belt also contains a number of asteroids, and simple frozen balls of water, ammonia and methane. The total mass of the Kuiper Belt is unlikely to weigh much more than 10 per cent the mass of Earth. Hence, there was not enough material for a large planet to emerge there.

The Oort Cloud begins approximately 27 light hours from the Sun, meaning light takes over a day to travel there, but it is still held by the Sun's gravity. It is composed of icy planetesimals and comets. It stretches onwards for an entire light year from the Sun. It might even extend as far as 3 light years from the Sun, nearly to our neighbouring star, Proxima Centauri, which is 4.2 light years away. This sphere of ice represents the very edges of our solar system, and the border between us and the rest of the galaxy.

Beyond our solar system, the Milky Way contains 200 to 400 billion stars. Many of these star systems also contain planets. We've found thousands of exo-planets (planets external to our solar system) in nearby solar systems by looking only at roughly 0.0000000000000000009 per cent of the total number of stars in the galaxy. In total, there must be trillions of planets in our galaxy. It is estimated that 300 million planets in the Milky Way may be capable of sustaining life just like Earth can. These are tremendously good odds that life has emerged elsewhere and that we are not alone in the Universe. Especially when one considers there are 400 billion to several trillion galaxies out there.

THE EARTH

As the thirty proto-planets of the early solar system continued to crash together in apocalyptic collisions, the resulting planets got bigger and bigger. Around 4.5 billion years ago, there were two planets in the orbital track where the Earth now is. I think you know where this is going . . .

One planet about the size of the Earth and another planet about the size of Mars, which we call Theia, smashed together. The Earth-sized planet absorbed most of the rubble and reformed. But 1.2 per cent of the matter drifted into Earth's orbit as wreckage from the collision. These shards cobbled together to become the Moon.

The Earth at this time remained extremely hot, as the fires of planetary collisions continued to burn. The Earth was also being pelted constantly by asteroids, each impact as devastating as a nuclear war. As the Earth continued to hoover up the surrounding matter in its orbit, the pressure from the weight of all that 'stuff' created heat in the Earth's core. In short, Earth 4.5 billion years ago was melted and squishy. A gelatinous ball of pudding which burned and bubbled at thousands of degrees.

This kicked off the process of differentiation. The Earth was a molten ball of squishy semi-liquid rocks, through which material could pass reasonably freely. Many of the heaviest elements like iron and gold sank through the scorching soup down to the very core of the Earth. The iron created a ball 3400 kilometres (2100 miles) thick at the core of the Earth, giving the planet its magnetic field.

Only tiny traces of these heavier elements remained trapped in the cooling Earth's crust. Which is why something like gold is so rare for humans to find. Yet if you were to somehow be able to burrow into the molten recesses of the mantle and core of the Earth, you would find enough gold to coat the entire Earth's surface, gilding the continents from sea to sea.

Lighter elements bubbled to the surface. To the top of this soup bubbled forth a crust of silicon (the majority share of Earth's chemical composition) and also aluminium, sodium and magnesium. The lightest of all elements, like carbon, oxygen, and hydrogen, were ejected as gases to form the Earth's early atmosphere.

Yet the cooling of the crust was frequently interrupted by asteroid impacts in the Late Heavy Bombardment. No sooner did the crust begin to congeal at the top of molten soup than more impacts destroyed the thin layer and heated the world up again. It was only around 4 billion years ago that the bombardments were over and the crust was able to fully solidify.

Even in this lava-soaked hell, higher complexity formed on our planet. Roughly 250 chemical combinations were possible as Theia crashed into the Earth. And over 1500 different chemicals existed by the time differentiation had finished its work.

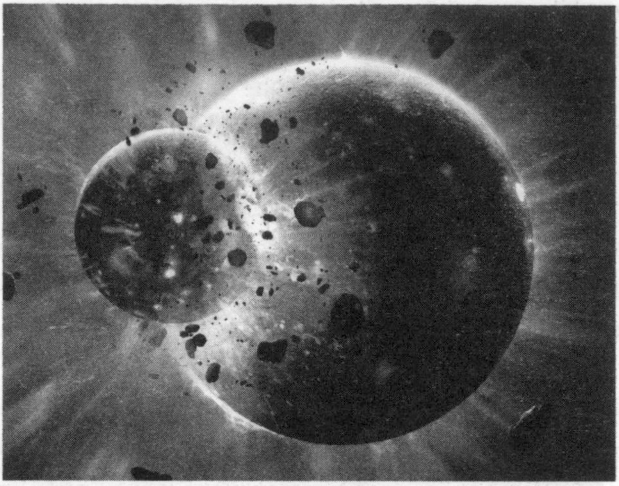

Earth and Theia colliding

TAKING MOTHER EARTH'S MEASUREMENTS

Even today, the crust is thin and frail by the standards of the rest of Earth's structure. This may be surprising when you consider how thick the rocky edifices of mountains and the dark, stony halls of mineshafts appear to be to human perception. But it is quite apt to call it the 'skin that forms at the top of a pot of soup'. The crust, which contains a lot of the Earth's lighter elements and just smidges of the heavier ones, is only about 35 kilometres (21.7 miles) thick and in some places at the bottom of the oceans about 7 kilometres (4.3 miles) thick.

Below the crust is the upper mantle, where there is so much pressure that temperatures increase to over 1000°C (1832°F), producing the horrific lava that occasionally spews to the surface from volcanoes. The upper mantle runs about 650 kilometres (403 miles) deep, in a sea of molten igneous rock. Below that is the lower mantle, which runs down to

2900 kilometres (1801 miles) deep, where it is so hot that rocks take a fully liquid form.

Further down is the core. The outer core is composed primarily of liquid iron and nickel, which flows down to 5200 kilometres (3231 miles) below the surface, and then the inner core, which runs down 6370 kilometres (3978 miles) to the very middle of this hell, and experiences so much pressure that the extremely hot molten core nevertheless behaves as if it were solid. At the core of the Earth, temperatures increase to 6700°C (12,000°F).

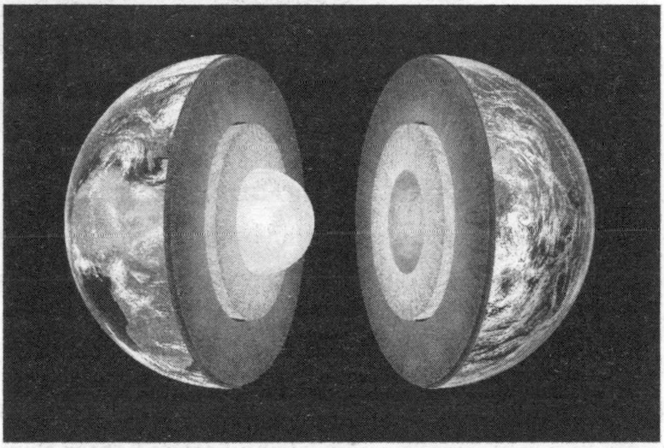

The Earth's crust, mantle and core

HELL ON EARTH

The Earth from 4.5 to 4 billion years ago belongs to the Hadean eon, so called because of the hellish conditions on Earth at the time. The surface of the Earth remained above 100°C (212°F), preventing any liquid water from forming, and in some places temperatures were as high as 1500°C (2732°F) and Earth was covered in oceans of lava.

Where there was land, it was paper-thin and jets of steam were shooting up out of its cracks as the lighter gases escaped the Earth in the process of differentiation. This job was also done by volcanoes, which rose out of the Earth, spewing lava, smoke and ash. As that lava dried into crust and built up, some volcanoes grew higher than Mount Everest.

The sky itself was a terrifying red. This was caused by the predominantly carbon dioxide atmosphere (about 80 per cent). It would be a long time before Earth adopted an oxygen atmosphere; for now, oxygen remained at negligible levels. And because the Sun was still young and not burning as brightly yet, this had the other haunting effect of making the sky fairly dark, aside from the redness. If there had been an 'eye of Sauron' hovering in the sky, like in *The Lord of the Rings*, it would not have looked entirely out of place.

When Theia smashed into the Earth approximately 4.51 billion years ago, a lot of the crust was destroyed and flung up into space along with an immense amount of lava. One cannot understate the catastrophic nature of a Mars-sized planet smashing into Earth; if it were to happen today, it would wipe out every trace of life and perhaps even evaporate all the oceans. It would have been roughly 450 times more severe than the asteroid impact that wiped out the dinosaurs 66 million years ago.

As the Moon slowly accreted and appeared in the sky, it was much closer to Earth than it is today (it moves away at about 4 centimetres a year) and would have blocked out much of the sky as it passed overhead. The tidal force exerted by the Moon would have been correspondingly greater. Leveraging massive tsunamis thousands of metres high to wash across the Earth every twelve to fifteen hours. Except these tsunamis were not made of water – they were made of molten lava.

But it gets worse. For 500 million years, the Earth was pelted by asteroids in the Great Bombardment, and particularly severely around 4.1 billion years ago in the Late Heavy Bombardment. This is where millions of asteroids struck the Earth, shattering its already thin and fragile crust. As Earth swept up the remaining cosmic debris in its region of the solar system, it suffered continual catastrophic impacts. Some of these would have been as powerful as a nuclear holocaust, some as bad as the Cretaceous impact that killed the dinosaurs, and some of them would have been up to 100 times worse (literally), depending on the size of the asteroid. And unlike the Cretaceous extinction, which happened once 66 million years ago, these impacts were happening constantly.

It goes without saying that any form of life would perish under these brutal conditions. There was no room for something as intricate as life in any place in the solar system at this point. Something on Earth needed to change for life, microscopic and frail, to stand a snowball's chance in hell of forming.

THE FIRST OCEANS

Despite the destruction, it appeared that roughly 500 million years of hell was quite enough. All that differentiation had belched hydrogen and oxygen into the atmosphere in a process called 'outgassing'. The bombardment of millions of asteroids had brought tonnes of ice from outer space. This promptly melted and rose into the atmosphere as well. Over time, the crust had cooled into a black and grey igneous landscape, with no more oceans of lava. And the surface temperature fell below 100°C (212°F) and continued falling. Suddenly all that water vapour that had built up in the atmosphere had no choice but to fall to Earth.

What followed was something that would resemble the Biblical flood. But never mind raining for forty days and forty nights; the torrential downpour across the planet continued for millions of years without end. The trenches and low-lying areas of the Earth's crust began to fill up with water. By approximately 4 billion years ago, the Earth was covered with oceans. Only the highest ledges, our continents, managed to keep their heads above water. And even these became pockmarked with lakes and streaked with rivers. Four billion years ago, the Hadean eon was over; the Archean eon had begun.

There are a few things to note about the Archean world. First, the Earth below the crust was still much hotter, since it had only recently formed, and emitted a lot of geothermal energy, which would be useful for the first forms of life. This compensated for the fact that the Sun's energy (solar thermal energy) was still quite dim, so was a less appealing power source for early life. But even if life had emerged on the surface, the solar radiation blasting down on the surface of an Earth without an ozone layer would have destroyed any life that tried to form. So, for now, the best bet for life was deep in the oceans, where it was warm and safe from radiation.

Archean Earth also had a gigantic Moon, which passed overhead, pulling tremendous tides from shore to shore. At least those waves were no longer lava. But the land itself was still dotted with an abundance of volcanoes as eruptions and outgassing continued. These volcanoes pumped out primarily carbon dioxide, which was still the dominant gas in the atmosphere. The other important thing to note about the land is that it was entirely rocky. All the greenery we associate with plains and forests did not exist yet. Instead, the Earth looked more like the surface of the Moon. Just add water.

And so the Inanimate Phase came to an end. The Archean Earth sat soulless and silent except for the sound of waves beating the rocky shore. It could have remained that way forever. If not for a highly unlikely event, our history would have already drawn to a close. For the next phase of complexity, we need to look to the bottom of the oceans. It is here we will find the first microscopic seeds of our family tree.

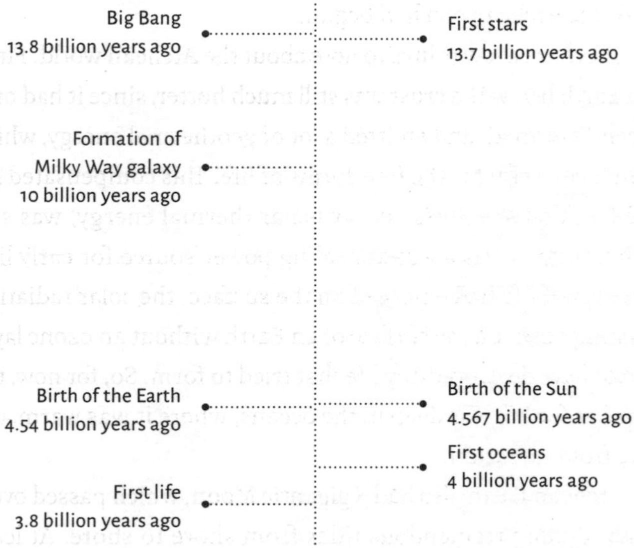
Big Bang
13.8 billion years ago
First stars
13.7 billion years ago
Formation of Milky Way galaxy
10 billion years ago
Birth of the Earth
4.54 billion years ago
Birth of the Sun
4.567 billion years ago
First oceans
4 billion years ago
First life
3.8 billion years ago

PART TWO

THE ANIMATE PHASE

3.8 billion to 315,000 years ago

4

LIFE AND EVOLUTION

Wherein the Earth becomes slightly less deadly, giving life a fighting chance • Differentiation and asteroid impacts create the world's first oceans • Within those oceans, long strands of organic chemicals begin to form • Those organic chemicals begin to self-replicate and evolve, triggering the existence of life • Some of that life becomes photosynthesisers • These photosynthesisers mess up the atmosphere and kill off a bunch of life • Against adversity, eukaryotes and sexual reproduction evolve • The final Snowball Earth event creates the first multi-celled life.

IN THE TRANQUIL SMOULDERING SEAS of the Archean Earth, 3.8 billion years ago, life began. We derive that date from chemical signatures left in Archean rocks by early microscopic life. By 3.5 billion years ago, we can actually see the fossil 'footprints' of these tiny microbes. Even this simple primitive life outstripped in complexity everything we have seen thus far.

By 4 billion years ago, the Earth's surface temperature had dropped below boiling point, and millions of years of rainfall created the first oceans. These were essential to life because it could not form if entombed in solid rock, where it could not move around. Or being fried by radiation on the Earth's surface. Or in the wispy gases of clouds. Liquid water was the ideal environment that allowed organic chemicals to move and join together in a soup-like mixture. Primitive life was fragile; it was a miracle it formed at all. The safest bet was to dwell at the bottom of the oceans.

But where would life get its energy flows, from which higher complexity could be created? The most likely answer is from undersea volcanoes or 'vents', which pumped out geothermal energy from cracks in the Earth's crust. Microbial life sat on the edges of these volcanoes, basking in the heat.

So we have the soup, we have the stove: now we just need the ingredients. The Archean oceans were teeming with a variety of organic chemicals that had bubbled to the surface via differentiation. It is no surprise that most of those organic chemicals, such as carbon (on which all terrestrial life is based), are among the lightest in the A periodic table. Carbon is also the most flexible. It forms a vital link in the chain for about 90 per cent of all the chemical combinations that we have discovered.

Besides carbon, equally vital to self-replicating life are hydrogen, oxygen, nitrogen and phosphorous. On the edges of undersea vents 3.8 billion years ago, the elements above came together to form complex organic chemicals: amino acids and nucleobases, long strings of building blocks.

Amino acids are crucial for fuelling life. You can find them in your food. They are a combination of carbon, hydrogen, oxygen and nitrogen atoms, wrapped in a chain of about nine atoms. Amino acids are the building blocks of proteins. Each protein is a tangled strand of about twenty amino acids on average, though some have considerably more. A protein is used to carry out the various commands of a living cell: to burn energy to sustain its complexity, to reproduce, to grow various traits, to react to its environment and also to simply move things around a cell.

Nucleobases, on the other hand, are the building blocks of nucleic acid (the fundamental component of DNA and RNA). These key chemicals are adenine ($C_{10}H_{12}O_5N_5P$), guanine ($C_{10}H_{12}O_6N_5P$), cytosine ($C_6H_{12}O_6N_3P$) and thymine ($C_{10}H_{13}O_7N_2P$). As you can see, we've come a long way in complexity since the first hydrogen atoms (H) at the start of the Universe.

DEOXYRIBONUCLEIC ACID: THE SEXIEST ACID OF THEM ALL

DNA exists in all living cells and is the database which tells proteins what sort of traits those cells should have and how they should behave. It is the 'software' of the organic computer, the disk which contains the program instructions that runs a video game. DNA is what makes living things look and act like they do. From fangs to freckles, from growling to laughing.

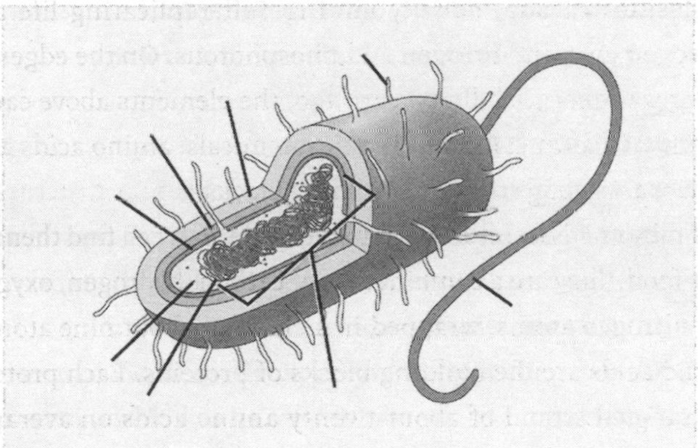

A prokaryotic cell

Deoxyribonucleic acid is made of two strands composed of billions of atoms, twisted around each other in a double-helix. Each strand is composed of many nucleotides, which in turn are composed of the nucleobases mentioned earlier, which may well have formed in the oceans of Archean Earth. Adenine, guanine, cytosine and thymine are the nucleobases that hold the genetic information. They are like the 1s and 0s encoded on a computer game disk.

Which brings us to the 'hardware' of our organic computer: ribonucleic acid, or RNA. Made of only one strand rather than two, RNA has the task of taking the instructions from DNA and delivering them to small parts in a living cell that produce proteins (these protein factories are called ribosomes). RNA does this by unzipping DNA and reading the instructions, or the 1s and 0s. The RNA then gives the proteins their marching orders. And those proteins set about building the organism. RNA and the proteins are essentially the disk reader and the hard microchip components of the computer itself.

3.8 billion years ago this highly structured, organic, acidic sludge started carrying out some highly intricate but seemingly arbitrary chemical reactions. But how did it evolve?

THE SOURCE OF EVOLUTION

How we went from basic organic chemicals to a structure as complex as DNA and RNA remains a blank page in our history. But once those structures were in place, their chemical reactions didn't happen just once.

DNA self-replicates, or copies itself, in order to continue giving instructions to the rest of a living cell. When it does so, it splits in two. Most of the time this copying process is perfect. But occasionally there is a copying error, or mutation, which slightly modifies DNA's instructions. A mutation happens maybe once in every billion copies. When the DNA mutates, it creates a slightly different organism.

If DNA copied itself flawlessly every single time without a single failure, life would have remained exactly as it was 3.8 billion years ago on the edges of undersea volcanoes. It would not have evolved. Mutations create historical change in biology.

Some mutations are deadly to the organism, some don't affect its survival one way or another and some mutations are useful. Those mutations that are useful are able to copy themselves all over again, and the cycle continues. Those mutations that work the best in a specific environment continue to exist. If not, those mutations (and the organisms possessing them) die out.

That is what evolution actually is: the natural selection of genes based on their evolutionary usefulness, rather than selection of the individual or the entire species. As environments change, so do the genes that work the best.

Thus, in this collection of organic sludge, we have all the key traits of a living organism: it uses energy flows from geothermal vents and surrounding amino acids (metabolism: that is, it eats), it reproduces by copying itself (reproduction) and it gradually changes its traits based on useful mutations (adaptation). These three traits of metabolism, reproduction and adaptation are the best idea we have when it comes to defining what the heck life is, and how it differs from the inanimate cosmos.

Once the process of self-replication and evolution began 3.8 billion years ago on the edge of underwater volcanoes, that organic sludge transformed into a diversity of strange new forms, eventually covering the whole Earth. Every bacterium, every plant, every animal and every human today is shaped from that 3.8-billion-year-old lump of clay. As Darwin wrote at

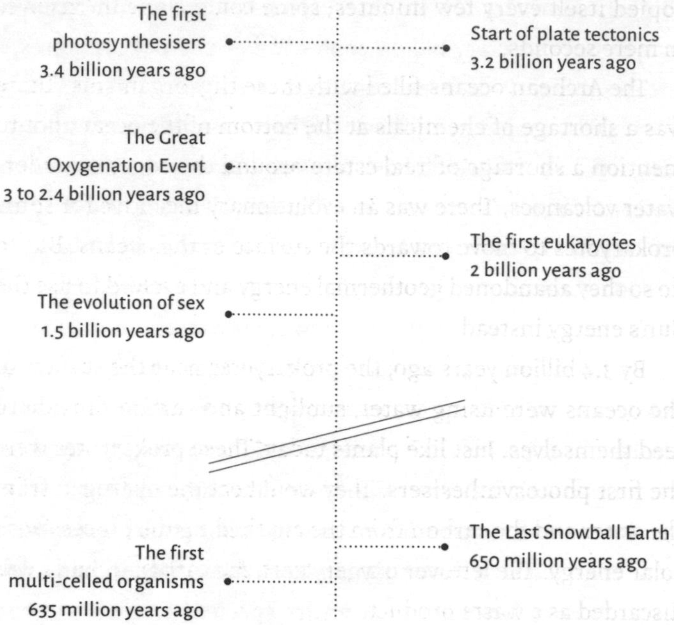
The first photosynthesisers
3.4 billion years ago
Start of plate tectonics
3.2 billion years ago
The Great Oxygenation Event
3 to 2.4 billion years ago
The first eukaryotes
2 billion years ago
The evolution of sex
1.5 billion years ago
The Last Snowball Earth
650 million years ago
The first multi-celled organisms
635 million years ago

the end of *On the Origin of Species*, 'From so simple a beginning, endless forms most beautiful and most wonderful have been, and are being, evolved.'

THE FIRST PHOTOSYNTHESISERS

The first organisms to emerge at the bottom of the oceans took in geothermal energy from undersea volcanoes and chomped down the chemicals around them. These organisms were, of course, quite simple. These were the 'prokaryotes': microscopic, single-celled organisms without a nucleus. Their DNA floated openly around the cell, increasing the risk of it getting damaged. These prokaryotes did not have sex (*quelle horreur!*) but cloned themselves, by splitting into copies. Each cell split and copied itself every few minutes; some could clone themselves in mere seconds.

The Archean oceans filled with these tiny organisms. There was a shortage of chemicals at the bottom of the ocean, not to mention a shortage of 'real estate' around the edges of underwater volcanoes. There was an evolutionary incentive for some prokaryotes to move towards the surface of the oceans. But to do so they abandoned geothermal energy and evolved to use the Sun's energy instead.

By 3.4 billion years ago, the prokaryotes near the surface of the oceans were using water, sunlight and carbon dioxide to feed themselves. Just like plants today. These prokaryotes were the first photosynthesisers. They would eat the hydrogen from the water and the carbon from the air, fuelling the process with solar energy. The leftover oxygen from the carbon dioxide was discarded as a waste product.

All these slight microbial changes took 400 million years. That's over 5 million average human lifespans. The same space of time that separates us from the first 'fish' that crawled out of the oceans and onto land . . .

Some photosynthesisers began to form large colonies: big mounds of microbes 50 to 100 centimetres in height, called stromatolites. Fossilised remains of these colonies can be found at Shark Bay in Western Australia today. They are roughly 3 billion years old.

THE OXYGEN HOLOCAUST

Even at this early stage, living things had the propensity to screw up the environment. I mentioned that the first photosynthesisers excreted oxygen as a waste product (O_2 left over from eating carbon dioxide). This was basically harmful 'poo' that photosynthesisers did not need. That is because O_2 is highly reactive, in the sense it creates violent chemical reactions. In large quantities, O_2 could kill the prokaryotes. Fortunately, 3.4 billion years ago, the amount of O_2 in the atmosphere was next to nothing.

But this slowly changed.

Around 3 billion years ago, the numerous photosynthesisers in the oceans were excreting so much O_2 it could no longer just be reabsorbed by rocks in the Earth's crust as it previously had been. So leftover O_2 entered the atmosphere. By 2.5 billion years ago, the level of oxygen in the atmosphere had increased from next-to-nothing to about 2.5 per cent. This was enough for organisms that had not evolved in an oxygen environment to suffer.

Scores and scores of prokaryotic species (all potential ancestors) died off. This was the single deadliest extinction event in Earth's history – albeit one that only affected microscopic single-celled organisms. And it is an extinction that living things blindly imposed upon themselves.

It should be noted that this was a slow death. A process that lasted roughly the same amount of time that separates humans from the Cambrian Explosion, or roughly 550 million years. Less complex life, tiny microbes, take longer to evolve and longer to impact the environment. Nevertheless, once felt, life's power is massive and irrevocable.

Small changes writ large in the Universe – a theme that will continue.

PLATE TECTONICS

From 3.8 to 3.2 billion years ago, the movement of melted rocks and lava below the Earth's surface created continual pressure upon a crust that was as thin as an eggshell relative to the

Earth's mantle and core. The immense pressure from all this red-hot movement created flashpoints where gigantic volcanoes exploded out of the Earth's surface. These giant volcanic plumes may have been responsible for 'cracking' the Earth's eggshell crust.

By 3.2 billion years ago, the regular, uninterrupted flow of plate tectonics began. The Earth's crust had been shattered into plates that were buffeted about by the flows of lava and squishy rock in the mantle below them. These movements are called convection flows. These push the plates about and constantly change the face of the Earth by moving continents, creating mountains, new oceans and a deluge of earthquakes and volcanic eruptions.

Imagine a saucepan of clam chowder on the stove. The cool air from the kitchen creates a skin on top of the soup. But liquid is bubbling away below it. If that bubbling gets too vigorous, it may 'break' the skin and propel different chunks of it around the top of the pot. That is plate tectonics in a nutshell. Or in a saucepan, at any rate.

THE OZONE LAYER AND THE FIRST SNOWBALL EARTH

The increase of oxygen in the atmosphere didn't stop 2.5 billion years ago. It accelerated. Levels of O_2 continued to rise as it gassed out of the oceans. Then, 2.2 billion years ago, it started to enter the upper atmosphere. The heat from the Sun began to transform the O_2 into O_3 in a process called photolysis. This is where the Sun knocks two oxygen atoms apart and then those single oxygen atoms combine with other molecules of O_2 to form O_3. A layer of O_3 began to blanket the Earth. This was the ozone layer. It reflected much of the Sun's rays, which had previously scorched the surface, back into space.

With little to counteract it, the blanket of ozone got thicker and thicker. With less of the Sun's heat reaching the surface of the Earth, the whole planet started to cool.

The oceans started to freeze at the Earth's poles. A thick layer of ice began to form. But it didn't stop there. The ice sheet began to descend from the poles of the Earth towards the equator. With each movement forward, the snow-covered white ice began to reflect even more of the Sun's rays back into space. This intensified and sped up the process of plummeting temperatures and the Earth's freezing. The average global temperature would have been around −50°C (−58°F). Eventually, two massive ice sheets, many metres tall, met at the equator and joined together, encasing the Earth in a tomb of ice. This time is known as 'Snowball Earth.'

THE RISE OF THE EUKARYOTES

From 2.5 to 2 billion years ago, some forms of microscopic life evolved the ability to use oxygen for energy. This process is called respiration. Instead of converting water and carbon dioxide into energy and giving off oxygen as a waste product, as photosynthesisers do, a respiring or 'aerobic' cell takes O_2 and gives off water and CO_2 as waste products. These microscopic single-celled organisms began gobbling up the oxygen in the atmosphere.

Two billion years ago, Snowball Earth placed a strain on all living species. The new oxygen-eaters had to be made of some pretty stern stuff to survive. They evolved into eukaryotes, with a radically more complex single-cell structure than a prokaryote. O_2 actually affords a cell much more energy once it evolves to digest it, so there was plenty of increased energy to fuel this evolution of a 'beefier' cell.

The eukaryotes were larger by about ten to 1000 times.

They were still microscopic, though the largest ones could almost be seen with the naked eye. Unlike in prokaryotes, their DNA was protected by a nucleus. The structure of the cell had a cytoskeleton holding it up (think about tent poles holding up a canvas tent). The eukaryotes were a damn tough domain of various species. They also represented a slight increase in structural and energy complexity. This allowed them to survive the Snowball Earth period.

Eventually volcanoes broke through the sheets of ice covering the Earth and began pumping carbon dioxide back into the atmosphere. This had the effect of warming the Earth. As the ice sheets receded, the carbon dioxide that had been trapped in the rocks of the surface and seabed also began to release CO_2 into the atmosphere. The cycle was reversing itself. The Snowball Earth phase was over – for now – and both aerobic and anaerobic species were able to flourish.

SEXY EUKARYOTES

After the retreat of Snowball Earth, the eukaryotes found a thousand new niches opening up for them. Some eukaryotes continued to respire oxygen, using a new organelle (miniature organs within a single-celled organism) called mitochondrion. Other eukaryotes evolved to become photosynthesisers, and instead of mitochondria had an organelle called the chloroplast. The former was the ancestor of animals, the latter was the ancestor of plants. We share at least 30 per cent of our DNA with everything in the plant branches of the family tree, whether it be daisies or bananas. We share even larger amounts of DNA with other animals.

Approximately 1.5 billion years ago, some catastrophe and period of ecological strain (the cause of which is unclear) led to a shortage of food for eukaryotes. Perhaps it was a regional crisis, perhaps it was global. But the lack of food began to cause eukaryotes to eat one another, surviving via cannibalism.

This act of cannibalism in a few cases must have resulted in the accidental exchange of DNA. In short, this Hannibal Lecter–esque act was the world's first glimpse of sex. Up until roughly 1.5 billion years ago, all eukaryotes simply cloned themselves like prokaryotes do. But now *some* eukaryotes had sex. The evolutionary advantages of sexual reproduction are profound. The exchange of DNA adds a greater amount of genetic diversity. The mutations of DNA are doubled in frequency, and the mixture of genes between two parent cells can also yield advantageous results. Evolution thus can move at a faster pace.

The first 'sexy eukaryotes' still divided like cells normally did. But instead of reproducing all their DNA in an exact copy, they would only reproduce half. The cell would then be tasked with finding a 'mate' to combine with, in order to complete the number of chromosomes required to create a new organism. Those single-celled organisms that did not find a 'mate' died off.

So advantageous was this process to evolution that it spawned a whole new range of tactics and behaviours and, eventually, instincts. Once organisms became multicellular, they began to compete for mates in ways that would influence the evolution of entire species' behaviour. The drive to have sex and reproduce became so engrained in organisms instinctually that it formed one of the primary motives for living: to survive long enough to attract a mate and reproduce. So powerful and pervasive is sex in evolution that it shaped an overwhelming majority of the traits of complex species and the overwhelming majority of their instincts (making organisms rather Freudian as a result), which in the case of humans bled into how we acted, rationalised and prioritised goals, and even how we shaped our cultures and societies.

THE LAST SNOWBALL EARTH (WE HOPE)

The trend of photosynthesisers pumping too much oxygen into the atmosphere repeated itself in the last billion years. This grew particularly severe when there was not as much volcanic activity to counterbalance the oxygen by pumping carbon dioxide into the atmosphere. As a result, the last billion years experienced two other phases of Snowball Earth. Not just an Ice Age, but the full encasement of the Earth in a layer of ice. There was one case of it approximately 700 million years ago, and another that began 650 million years ago and ended 635 million years ago.

The final Snowball Earth imposed yet another period of strain on Earth. Those eukaryotes that had evolved to sexually reproduce were able to adapt more quickly to the harsher conditions. Some of these 'sexy eukaryotes' came to live in colonies where different microbes filled different functions in a form of symbiosis, which allowed everyone in the colony to survive the freezing conditions.

The last Snowball Earth sent symbiosis into overdrive. Eukaryotes weren't just living symbiotically in colonies anymore. Each individual set of microbes had become so specialised in their duties in the colony that one kind of eukaryote could not live without the other. Thus, with the pressures of the last Snowball Earth, the first multi-celled organisms (the ancestors of plants, animals and fungi) were born.

MULTICELLULARISM

There comes a point where symbiosis between single-celled organisms becomes so intense it crosses the threshold into multicellularism. For example, you don't *just* have symbiosis with your liver cells. Your liver cannot crawl along the ground behind you when you go to the shops. It is such an inextricable part of you that you are in fact one structure, one organism.

A multicellular organism is a collection of multiple trillions of cells, each of which is shaped by DNA to act differently, fill a function and coalesce with similar cells to form an organ. The organs themselves are built up into a patchwork of intricate networks, such as the circulatory system, the respiratory system and the digestive system.

To give you an idea of the scale of difference, there are 37 trillion cells in a single human body. There are only roughly 400 billion stars in the Milky Way. Thus, there are approximately 92.5 galaxies of single cells in one human body. Complexity in terms of building blocks and intricate structures has gone beyond anything that has been seen so far in our story.

A multicellular organism has a lot more moving parts that can break down. So it is not like there is always an evolutionary incentive for living things to become more complex like this. With greater complexity comes greater fragility. It is why the majority of life on Earth is still single-celled. It is only when a species is *forced* by the environment to evolve into becoming more complex that it does so.

As a wider rule, the same logic is why the majority of the Universe is fairly simple and most atomic matter is hydrogen. Complexity is in many ways the exception, not the rule. All stretching back to those tiny pinpricks of unequal energy appearing a split second after the Big Bang in a Universe that was already 99.9999999999999 per cent equally distributed energy – or dead.

BIOLOGICAL COMPLEXITY

In order to create, sustain or increase complexity, you need energy flows from where there is more energy to where there is less. For an organism to sustain its own complexity and stave off death, it requires greater density of energy flow (for its size) than something like a star:

- the Sun: 2 erg/g/s (a unit of free energy per gram per second)
- A typical microscopic organism: 900 erg/g/s
- a tree: 10,000 erg/g/s
- a dog: 20,000 erg/g/s.

While a microscopic bit of organic sludge 3.8 billion years ago would not have been as imposing as a star (it is microscopic and extremely fragile, after all), it requires a great deal more energy for a single cell to maintain all its working parts.

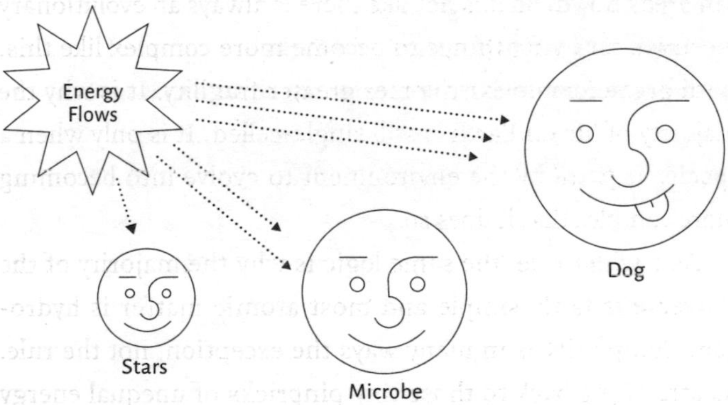
Energy Flows
Dog
Stars
Microbe

From the first inequalities in matter and energy that emerged after the Big Bang, to stars, to planets and now to organisms, tiny pockets of complexity are growing brighter and brighter in terms of their energy density. This trend will continue for the rest of our story.

But how is life to meet the increased demand for energy flows? Simple: it must go out and *actively* find them. While a star is content to float in space for billions of years and simply burn up its fuel, a living being must *actively* seek out new energy flows in order to sustain its own survival. It does this by chemosynthesising, photosynthesising, munching on plants, hunting or making a trip down to McDonald's at 2 a.m. after too many beers. You don't see stars floating round the cosmos, hungrily chasing after fleeing clouds of helium. To actively seek out energy is one of the defining traits of living things.

It also means that at this point in our story, we have a growing sense of historical agency. We are no longer consigned to a passive, inanimate Universe that quietly awaits its fate. We have acquired the ability to grow, to change, to innovate and, where possible, to stave off our demise. Complexity no longer goes gently into that good night.

From this point forward, we fight for survival. And the greater the complexity, the greater the chance that we can win.

5

EXPLOSIONS AND EXTINCTIONS

Multi-celled life proliferates in the oceans • Eyes, the spine and the brain evolve • Plants, then bugs, then vertebrates slowly venture onto land • Extinction events followed by swift evolution create strange and bizarre new species • Complexity plateaus, with boom-bust cycles of Darwinian 'tooth and claw'.

WE NOW ARRIVE AT A MORE CLASSICAL VISION of evolution. That of 'nature, red in tooth and claw' where multi-celled organisms evolve and fight for survival. This stage represents an unprecedented level of complexity. While previous changes in our story from the Big Bang to now have been measured in the billions or hundreds of millions of years, evolutionary changes are going to come a lot faster. That is another side effect of higher complexity. Faster rates of change. And more profound impacts on the surrounding environment. In that respect, the last 635 million years have been very eventful indeed . . .

The period of time 635 million to 66 million years ago is primarily characterised by the pattern of explosions and extinctions. An explosion of new evolution as life develops revolutionary new traits that open up thousands of new niches in the environment. Or life rapidly filling old niches left empty by some catastrophic extinction event that killed a large percentage of species on Earth. Each time these things happened, we see the appearance of something new.

But it is important to note that none of that was written in the stars. We could very easily never have evolved. And the entire experiment of life on Earth could have ended with one well-placed asteroid hundreds of millions of years ago. When I consider the hundreds of millions of years of evolution where something different could have happened, or the thousands of human ancestors who might never have survived, or when I add up all those probabilities, in addition to being the one sperm of billions that made it, the fact that I get to exist at all in this Universe makes me feel extremely lucky.

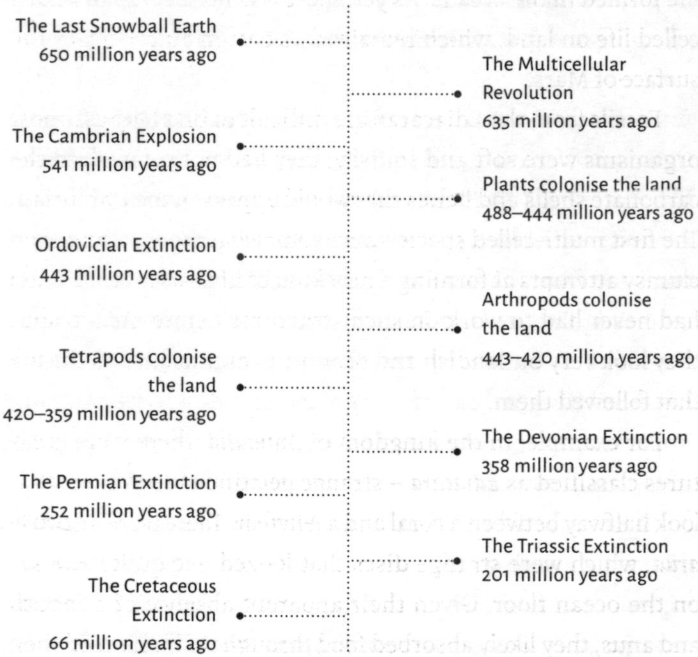
The Last Snowball Earth
650 million years ago
The Multicellular Revolution
635 million years ago
The Cambrian Explosion
541 million years ago
Plants colonise the land
488–444 million years ago
Ordovician Extinction
443 million years ago
Arthropods colonise the land
443–420 million years ago
Tetrapods colonise the land
420–359 million years ago
The Devonian Extinction
358 million years ago
The Permian Extinction
252 million years ago
The Triassic Extinction
201 million years ago
The Cretaceous Extinction
66 million years ago

The Darwinian world is by definition a cruel one. Extinction is a necessary component of evolution. In order for the useful traits of an organism to be 'selected for' by natural selection, a legion of other competing organisms must cease to exist. There are only so many niches and resources in an environment. 99.9 per cent of all species that ever existed have gone extinct. 'Natural selection' is a bit of a misnomer; nature doesn't actively select so much as eliminate everything else.

We are the ones who survived.

THE EDIACARAN (635–541 MILLION YEARS AGO)

After the last Snowball Earth receded, oxygen levels in the atmosphere dropped thanks to volcanoes pumping out CO_2. The climate dramatically warmed as a result. The first multicellular life formed in the oceans. As yet, there was not a scrap of multi-celled life on land, which remained as barren and rocky as the surface of Mars.

Fossils from the Ediacaran are difficult to find because most organisms were soft and squishy; they had not yet evolved the carbonate shells and bones that would appear in the Cambrian. The first multi-celled species were somewhat modest, and even clumsy attempts at forming a new kind of life. Natural selection had never had to work on such structures before. As a result, they look very outlandish and bear little resemblance to the life that followed them.

For example, in the kingdom of *Animalia*, there were creatures classified as *Ediacara* – strange gelatinous structures that look halfway between a coral and a jellyfish. There were also *Arkarua*, which were strange discs that looked like quilts and sat on the ocean floor. Given their apparent absence of a mouth and anus, they likely absorbed food through their skin and then excreted waste back out in the same way. Then there was *Pteridinium*, which looked like a primitive worm. And *Charnia*, which looked like a long underwater fern. Almost all Ediacaran animals didn't have a means of locomotion, though some may have drifted across the ocean floor 'grazing' on what food they could find. It was a weird time. H.P. Lovecraft would have been a fan of the Ediacaran.

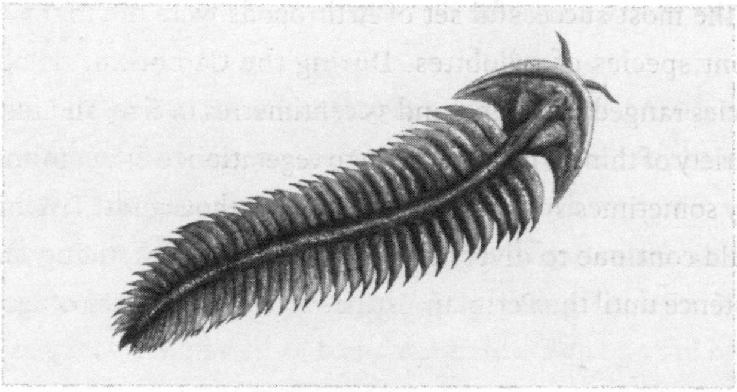

Ediacara

THE CAMBRIAN (541–485 MILLION YEARS AGO)

The Cambrian Explosion was a time of very rapid evolution, with multi-celled species entering new ecological niches. This process began 541 million years ago and lasted for roughly 15 million years. There are abundant fossils from this time due to the evolution of hard exoskeletons and shells. These were the Arthropods, the ancestors of crabs, lobsters, insects, arachnids and so on. Either creepy things or the most expensive thing on the menu.

The eye evolved. At first the eye was a primitive sensory tool that animals used to detect changes in light and movement.

The innovation stuck, which is why we see eyes across all animals, even when newly evolving species have a declining use for them, such as bats, badgers or deepwater fish. Eyes did not evolve in the same direction. For instance, many molluscs have eyes along their bodies rather than centrally located in what we would identify as a head. The eyes of ants or spiders are also vastly different to our own. Even in species as closely related as humans and dogs, they each have different aptitudes for sight.

The most successful set of Arthropods were the many different species of trilobites. During the Cambrian, trilobite species ranged between 5 and 35 centimetres in size, and fed on a variety of things, from bacteria to vegetation to other animals. They sometimes swarmed in hundreds or thousands. Trilobites would continue to diversify and would manage to hang on to existence until the Permian Extinction 252 million years ago.

Trilobite

The Chordates (our ancestors) had a somewhat humbler start. The first ones evolved 530 million years ago. They started as *Pikaia*, which resembled a worm and swam like an eel. *Pikaia* was only a few centimetres long. It had a single rod made from cartilage running along its body. A primitive spine. This was the ancestor of vertebrates. Due to its mode of swimming, one end of the species was always facing the front, to encounter food or danger. This led to cephalisation, the process by which sensory organs move increasingly to one part of the body, as such nerves began to make their way along the cartilage to what we would now consider a head. These are the first baby steps of the evolution of the brain. The trend of cephalisation continued 525 million years ago with the evolution of *Haikouichthys*, one of the first identifiable Cambrian jawless 'fish'.

Another innovation of the Cambrian era was predation. *Anomalocaris*, roughly 515 to 520 million years ago, was a vicious Arthropod approximately 1 metre in length (dwarfing most other Cambrian life in size). It had an armoured exoskeleton, and at the front it had two massive, grasping spiked claws. With these, *Anomalocaris* would scoop up unwitting prey in the oceans, impaling it on the spikes, and then bring it up towards its downward-facing mouth to be devoured. Hilariously, the name *Anomalocaris* comes from the roughly translated Latin for 'weird shrimp' or 'weird sea crab'.

In many ways, predation is just an inevitable evolutionary extension of energy flows in nature. If you can eat solar energy, eat chemicals, eat plants or eat dead things (as fungi do), why can you not evolve to eat the multi-celled creatures that eat all those things? The reason why the impaling and devouring by *Anomalocaris* looks much more heinous to humans is because we are instinctually driven to avoid being eaten ourselves. The processes of energy flows that we choose to dub either harmless or heinous are made subjectively, shaped by our evolutionary instincts and perspectives. This subjectivity somewhat complicates moral discussions between human omnivores and herbivores and our perspective of the innate cruelty of the Darwinian world.

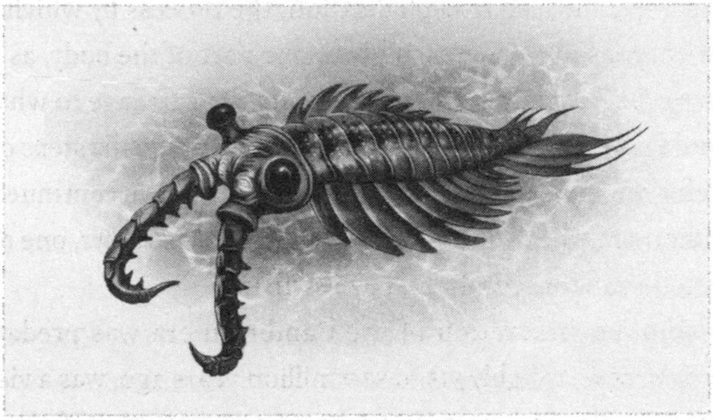

Anomalocaris

Predation set off an evolutionary arms race. In response to predators such as *Anomalocaris*, certain species of trilobites began to develop spikes on their exoskeletons in order to discourage others from eating them; other trilobites learned to roll up into balls to protect themselves. Or they developed camouflage and faster locomotion to avoid detection and danger. Other trilobites began to eat worms, jellies and other unprotected animals that were becoming predators themselves. The evolutionary arms race between predator and prey has continued to this very day.

THE ORDOVICIAN (485–444 MILLION YEARS AGO)

The Ordovician atmosphere was glutted with ten times the CO_2 we have today. In the early Ordovician, the average temperature of the ocean was somewhere between bath temperature and a hot tub (35° and 40° Celsius respectively). By 460 million years ago, the seas had cooled to an average temperature of 25° to 30° Celsius, which is still warm, about the temperature of tropical waters.

The first ancestors of octopuses and starfish emerged. Coral reefs formed in the warm waters. The ancestors of oysters, clams and sea snails all multiplied. The first sea scorpions emerged, some the size of modern scorpions, some the length of your shin. All told, the number of marine species multiplied by four times in the Ordovician compared to the Cambrian.

Meanwhile, the first multi-celled life began to venture onto land. These were the plants. Starting as very simple algae on coastlines and rivers, some of this algae evolved into little weed-like structures that were no more than 10 centimetres in height. The plants also acted in symbiosis with fungi, which furnished plants with minerals, and became closely attached to their roots.

Mass extinction loomed for the first time since Snowball Earth. Thanks to terrestrial plants, the levels of oxygen on Earth began to increase again, causing a cooling period that killed off warm-water species, but it was short-lived. CO_2 rapidly reached its old levels in the atmosphere, heating the world and killing off the species that had evolved to adapt to the cooler conditions.

In total, 70 per cent of marine organisms were annihilated. Their niches were left empty for the survivors to evolve and fill.

THE SILURIAN (444–420 MILLION YEARS AGO)

Plants continued their march inland, forming tiny shrubs and mosses. The majority of the Earth was still rocky, with just a smattering of pygmy forests near water sources.

Fungi progressed faster on land. Some of them grew to the height of several metres. While plant roots were primitive and could not yet break into the rocky exterior of the Earth, fungi literally ate into the rocks and flourished during this period.

In the seas, some fish evolved jaws and their spines became more articulated. Jawed fish soon led to the evolution of the first sharks, and the evolutionary arms race continued with other fish developing quicker reflexes and more complex brains.

Silurian jawed fish

Arthropods (bugs, lobsters and so on) came onto the land in the Silurian. Driven out of the seas by the pressures of the Ordovician extinction event, the first terrestrial Arthropods discovered a food resource in both dead and living plants. For instance, *Pneumodesmus* lived 428 million years ago and was an ancient millipede about 1 centimetre in length that fed off dead plant material. Not long behind these vegetarian bugs were arthropod predators, most notably the first spider-like arachnids.

The oxygen levels of the Silurian remained quite low, averaging about 15 per cent, so these predators remained quite small: only a few centimetres themselves. The Silurian was a world of tiny bugs and tiny plants in a realm dominated by fungus. A somewhat unappetising image.

THE DEVONIAN (420–358 MILLION YEARS AGO)

The world was temperate, probably contained little to no polar ice, and was lush and tropical over most of its surface, with the exception of deserts forming at the equator.

Fungi started forming towers and mounds that were up to 10 metres high and created increasing amounts of soft soils, which plants could pierce with their roots. As such, ferns and mosses spread into abundance beyond the riverbeds and the Earth finally became a lustrous green.

By 410 million years ago, some plants managed to grow up to 14 metres high, and by 380 million years ago, some plant species had evolved wood to strengthen their stems so they could maintain these massive heights or grow even higher as they competed for sunlight. Ergo, the first true forests.

There was a tremendous diversification of species in the oceans. Fish began to grow larger and sturdier, with some reaching between 3 and 7 metres in length. They developed ray and lobe fins, and more intricate body structures. Sharks became extremely numerous. Sea scorpions grew to a massive 2.5 metres long.

Devonian sea scorpion

Spiders began to develop their ability to cast silk webs in order to catch prey. Flying Arthropods appeared during this time and began exploiting the advantages of such mobility. The buzzing began.

The most profound change in the Devonian was the advent of *Tetrapoda* (or the first vertebrates) on the land. Our ancestors. 380 million years ago, the first lungfish emerged. They had a hole in the top of their head that was angled so that air could flow into primitive lungs. The first lungfish had strong front fins that it could use to drag itself along the bottom of shallow waters in pursuit of food. Gradually this impulse transitioned to dragging itself along beaches. By 375 million years ago, *Tiktaalik* was breathing air and had strong front and back fins and also primitive hips that would aid with locomotion.

An early tetrapod

By 370 million years ago, we had transitioned to the stem-tetrapods, such as *Ichthyostega*. About 1 to 1.5 metres in length, this was the first proto-amphibian, swimming in shallow swamps. The skull hole of its ancestors had evolved into nostrils. It had the signature four limbs and five digits of the earliest tetrapods. The same number of limbs and digits are present or 'vestigial' to all terrestrial vertebrates. This includes you, frogs, dogs, cats, horses, lizards, bears and even snakes. In the extreme case of snakes, these vestigial limbs can still be found – having shrunk to be almost imperceptible.

By the end of the Devonian, plants pumped out too much oxygen, cooling and drying out the planet. Amphibians (the only tetrapods) dried out and died. Roughly 95 to 97 per cent of them were wiped out. It is astounding to think that the diversity of tetrapod life on Earth today, from salamanders to owls to humans, are descended from a 3 to 5 per cent bottleneck. Meanwhile, the climate changes killed off about 50 per cent of aquatic life.

THE CARBONIFEROUS (358–298 MILLION YEARS AGO)

Giant carboniferous trees made the oxygen levels reach 35 per cent (today's oxygen levels are 21 per cent). The Earth was covered with carboniferous forests. Some of the trees grew to 50 metres in height. Because they had pumped so much oxygen into the atmosphere, they were the cause of their own demise. Horrific forest fires were exceedingly common. Large tracts of land dried out and forests could not grow anymore. And this left layer after layer of dead trees, which formed some of the massive coal beds we use today.

The increased oxygen created larger Arthropods. We're talking giant dragonflies with a metre-long wingspan, giant land scorpions nearly two metres long, giant ground-spiders, giant cockroaches and giant millipedes two metres long and half a metre wide. The Carboniferous would be an excellent setting for a time-travel horror film.

The first reptiles evolved between 350 and 310 million years ago, and their evolution was intensified after the carboniferous forests collapsed from a drying climate. Reptiles had tough skin that did not lose as much water. This meant they could head further inland, away from abundant water sources. Some could even survive in desert climates, which were slowly becoming more abundant. They started laying hard-shell eggs, which meant they did not need to return to the water to reproduce.

Giant insects during the Carboniferous

THE PERMIAN (298–252 MILLION YEARS AGO)

The oxygen level shrank to 23 per cent, shrinking the size of bugs. The larger kinds required more oxygen to survive. The most successful Arthropods in the Permian were the ancestors of cockroaches, which formed the overwhelming majority of insect biota in the period. Deserts filled with roaches. Yuck.

Reptiles flourished. The ancestors of mammals and dinosaurs in the Permian were *Synapsids* and *Sauropsids*, respectively. The *Synapsids* were proto-mammals that still looked very reptilian. They used mammary glands to rear their young. Many of them ate roaches. There is no accounting for taste.

The *Therapsids* evolved from *Synapsids*. They were energetic and fast-moving, and as a result they had a higher body temperature. In other words, they were warm-blooded. In order to maintain this temperature, many of them began to develop fur. 260 million years ago, a smaller group evolved from the *Therapsids*: the *Cynodonts*, which were small, timid creatures, many of which were capable of burrowing.

Therapsid

On the other side of the family tree, *Sauropsids* retained what we would describe as much more reptilian characteristics. These creatures were the ancestors of everything from turtles to crocodiles, to archosaurs, pterosaurs, dinosaurs and the birds (avian dinosaurs).

The Permian Extinction, or Great Dying, 252 million years ago, is theorised to have been caused by a volcanic super-eruption in what is today Siberia. It was a catastrophe that lasted for about a million years. Ash was thrown into the atmosphere, blocking out the sun and killing off vegetation. Acid rain hurtled down from the skies. Oxygen was stripped from the oceans. It was the one mass extinction event which nearly ended all complex life on Earth. This had such a traumatic effect on the planet that roughly 90 to 95 per cent of all species died out. The *Cynodonts* managed to survive because of their small size and burrowing behaviour. The surviving *Sauropsids* thrived in the new climate and were soon to rule the Earth.

THE TRIASSIC (252–201 MILLION YEARS AGO)

It was halfway through the Triassic before the biosphere recovered from the devastation of the Great Dying. It was generally quite dry, with massive deserts forming in the interior of Pangea, more arid than even in the Permian era. Rain simply could not reach the interior of the super-continent.

The *Sauropsid* line had spawned the archosaurs, from which all dinosaurs, pterosaurs and crocodilians are descended. The archosaurs had an advantage over other reptiles in that they had multiple lungs that allowed them to breathe in an atmosphere with only 16 per cent oxygen. Dinosaurs at the start of the Triassic were only a small minority. About 5 per cent.

234 million years ago, volcanism increased the climate and humidity of Earth. Rain was suddenly everywhere. In this 'pluvial episode', rain belted down across the Earth for 2 million years straight. This had a devastating effect on animals that liked the arid desert climates. Meanwhile, dinosaurs thrived in the more humid environments that emerged. The first pterosaurs began to take flight.

The Triassic extinction event 201 million years ago (its cause is unclear – but probably an asteroid) wiped out a great many amphibians, *Therapsids* and most archosaur species other than dinosaurs and pterosaurs. As a result, dinosaurs came to represent 90 per cent of all tetrapods on Earth. Proto-mammals cowered on the fringes.

THE JURASSIC (201–145 MILLION YEARS AGO)

The Jurassic opened with the super-continent Pangea cracking into pieces and the climate becoming increasingly humid. The modern continents began to take form, with North America and Europe merged together and South America and Africa still connecting their perfect 'puzzle pieces'. The gulf was increasing between these two continental couples. As a result, there was no longer a huge desert interior. Rainfall hit more of the land, which increased the amount of forest and heavy vegetation. The oxygen levels increased to approximately 25 per cent.

Dinosaurs filled the empty niches left by the Triassic extinction. The humid rainforests afforded herbivores a great deal of food, and the dinosaurs evolved to eat increasing quantities of vegetation. Thus, we see the evolution of species like

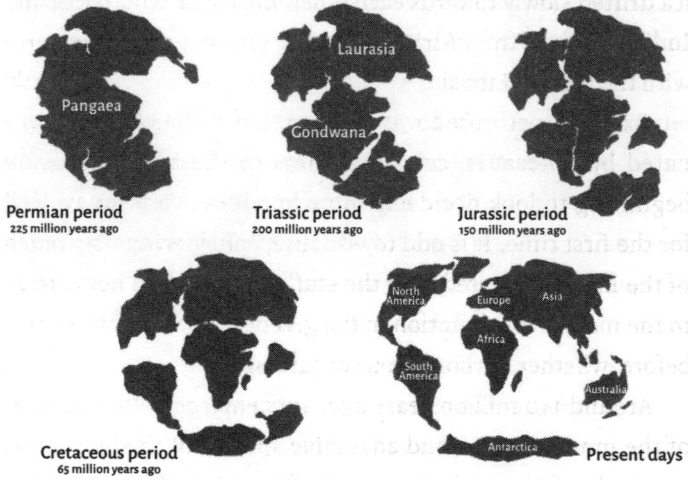
Pangaea
Permian period
225 million years ago
Laurasia
Gondwana
Triassic period
200 million years ago
Jurassic period
150 million years ago
Cretaceous period
65 million years ago
North America
Europe
Asia
Africa
South America
Australia
Antarctica
Present days

Supersaurus, which was 35 metres long, and *Allosaurus* (10 metres in length and a classical-looking dino predator) dominating the food chain.

Proto-mammals kept out of the way. Their average size was not much bigger than a mouse. They burrowed or hid up trees, ate insects and only came out at night. By 165 million years ago, a few of them became tree-bound and developed gliding abilities, and a few returned to the coastlines and habitats closely attached to water.

By the Late Jurassic, the first avian dinosaurs (the ancestors of birds) began to take flight. A sort of fuzz appeared on just a few Triassic dinosaurs for warmth and this now became feathers. Some dinosaurs remained coated with primitive feathers (even T. rex in the Cretaceous may have had fuzz), some dinosaurs had no feathers at all, but in some species feathers spawned flight.

THE CRETACEOUS (145–66 MILLION YEARS AGO)

The break-up of Pangea was complete. North and South America drifted slowly towards each other. Australia, Antarctica and India broke off from Africa, the latter set on a collision course with the belly of Eurasia.

Oxygen levels rose to 30 per cent. On an Earth still dominated by dinosaurs, certain corners of the biosphere were beginning to look decidedly more 'modern'. Grasses evolved for the first time. It is odd to visualise, considering how much of the Earth is covered with the stuff, but they had not existed in the masses of vegetation in the greenest phases of the Earth before, whether Carboniferous or Jurassic.

Around 140 million years ago, ants emerged. They are one of the most common and adaptable species of the biosphere, accounting for roughly 20 per cent of the Earth's biomass today.

Then, 125 million years ago, flowering plants (which had not existed before) evolved and spread across the Earth, largely thanks to their concurrent evolution with bees.

Around the same time, the first proto-placental and proto-marsupial mammals appeared on the fossil record. Both gave birth rather than laid eggs, with the former species gestating the offspring for longer in the womb and the latter giving birth and nurturing the offspring in a pouch or pocket. While still quite small and skittish, the placentals would become prominent in the Americas, Eurasia and Africa, and the marsupials would predominate in Australia. And the ancestors of the platypus laid eggs, confusing the hell out of everyone.

Meanwhile, the dinosaurs continued to reign supreme, with most niches being filled by them. This glut of dinos stiffened competition between species – particularly in terms of the balance between herbivores and the predatory carnivores that hunted them. As a result, you get some of the most astounding forms on both sides in this period – from the emergence of apex predators such as *Tyrannosaurus rex* and *Albertosaurus*, to the increasingly diverse defensive tusks used by *ceratopsia* like *Triceratops*, the long spines that sauropods grew on their necks to ward of predators, as with *Amargasaurus*, or the heavy armoured plating worn by *Ankylosaurus*.

An extinction event during the Cretaceous obliterated 70 per cent of the remaining species on Earth, including 90 per cent of terrestrial animals and 50 per cent of plant species. An asteroid 10 kilometres (6.2 miles) in diameter struck the Yucatán Peninsula. Worldwide earthquakes, tsunamis, continent-wide forest fires and huge torrents of acid rain killed off much life. Then all the dust flung into the air blocked out solar energy and killed more plant life, causing the surviving herbivores to starve,

Ankylosaurus

followed by the carnivores. The Earth became littered with rotting plants and animals, fed upon by flies, maggots and other corpse-feeders. Those birds and mammals that could eat insects and survive on what little plant life remained escaped the catastrophe, while non-avian dinosaurs perished. The niches had been swept clean once again, and this time it would be the mammals that would fill them.

The period of multi-celled organisms from 635 to 66 million years ago only saw slight increases in complexity overall. In fact, during this period complexity more or less plateaued. The Darwinian game of evolution and extinction might have been the height of complexity in the known Universe. Except by a series of accidents in a world dominated by mammals, a new and faster form of evolution would emerge, capable of achieving even higher levels of complexity: culture.

6

PRIMATE EVOLUTION

Wherein mammals fill the niches left empty by the dinosaurs • Primates evolve bearing decidedly familiar traits • Humans split off from our last common ancestors with the Great Apes • We start walking on two legs and our brains get bigger • We start accumulating more information with each generation than is lost by the next.

66 MILLION YEARS AGO, the world was a wasteland. Chilly and dry. The landscape was littered with dead plants and animals, rotting away in the sun and gradually being covered by layers of dust and dirt. Large species had been heavily hit in the Cretaceous extinction event, because of the collapse of the food chain. Aside from turtles and crocodiles, the land was now populated by tiny creatures like birds and mammals.

The boom-bust cycles of biological evolution continued as they always had. The ecological niches of the world had been swept clean. A rapid evolution of mammals filled them again. At first, mammals resembled mice or chipmunks. They were mostly under 50 centimetres long and weighed under a kilogram. They nibbled on plants and ate bugs. They burrowed into the ground to hide or concealed themselves in trees. These mammal survivors would diversify and dominate the Earth as the dinosaurs, *Sauropsids*, amphibians and Arthropods had done before them. And the vicious Darwinian cycle might have continued without any sign of increasing complexity for hundreds of millions more years. But this time there was something new on the horizon …

By 60 million years ago, the climate had warmed up again. The world was warm. North America and Eurasia were tropical. The majority of the Earth was covered in forests, with deserts at the equator. At the poles, there was little or no ice. And the mammals had begun to grow.

The ancestor of elephants at this time was no larger than a dog, but it would gradually evolve to become the world's largest

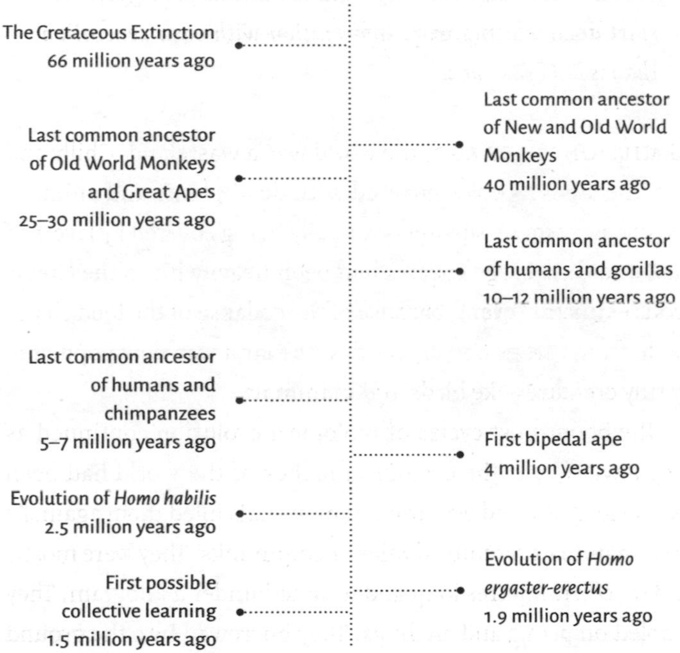
The Cretaceous Extinction
66 million years ago
Last common ancestor
of New and Old World
Monkeys
40 million years ago
Last common ancestor
of Old World Monkeys
and Great Apes
25–30 million years ago
Last common ancestor
of humans and gorillas
10–12 million years ago
Last common ancestor
of humans and
chimpanzees
5–7 million years ago
First bipedal ape
4 million years ago
Evolution of *Homo habilis*
2.5 million years ago
Evolution of *Homo
ergaster-erectus*
1.9 million years ago
First possible
collective learning
1.5 million years ago

land mammal. At the same time, a similarly sized mammal had begun to hunt fish and red meat, using sharp teeth for tearing the flesh of prey. By 42 million years ago, these predators had evolved into two branches with canine and feline characteristics: the ancestors of wolves, foxes and bears or lions, tigers and jaguars, respectively.

55 million years ago, a small mammal the size of a cat evolved to periodically spend time in the water and even submerge itself. This mammal was the ancestor of hippos and whales. The ancestors of whales began to spend more and more time in the oceans, first in shallow waters, later being able to dive deeper and eat scores and scores of krill and fish. By 40 million years ago, their transformation into whales was complete.

Also 55 million years ago, the ancestor of the horse – multi-toed and around the size of a dog – dwelt in the forests. It crept quietly and nimbly among the trees and the brush on the forest floor.

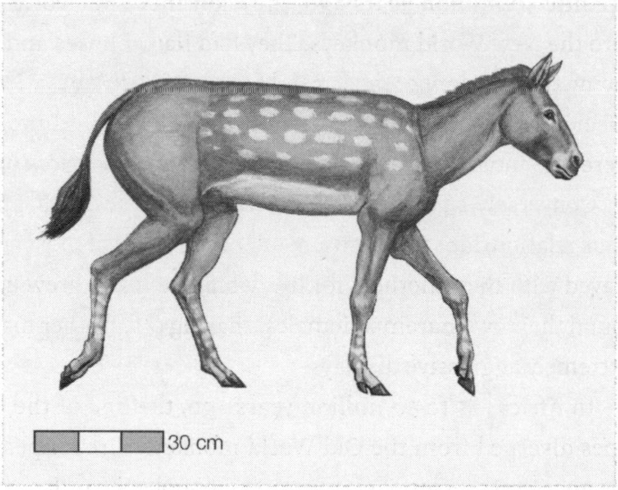
30 cm

Ancestral horse

Once the climate cooled and dried, these creatures began running increasingly on the dominant third toe. Over time, the other toes receded significantly, giving horses their characteristic hooves. They no longer crept through the forest but travelled vast distances.

In a relatively short amount of time – a few dozen million years – mammals rapidly filled environmental niches and grew from tiny sizes to form the bulk of the world's megafauna and the building blocks for the familiar species we know today.

Fifty-five million years ago was also when the primates emerged. They began as small tree-dwelling mammals with grasping hands and front-facing eyes. These traits were particularly useful to avoid tumbling from the trees. The front-facing eyes, for instance, permitted primates stereoscopic vision and depth perception, which is particularly crucial when judging a leap from one branch to another. In order to process all that 3D information, primates required increasingly larger brains.

Primates colonised the Americas and, separated by the vast Atlantic 40 million years ago, they continued to evolve there into the New World monkeys. They had flatter noses and side-facing nostrils, longer tails useful for grasping things, and most species do not have opposable thumbs. New World monkeys were also more likely to observe monogamous relationships.

Conversely, in most species of Old World monkeys, polygynous relationships were most common. Females of most species stayed with their mothers for life, while the males grew up and found their own harem of females, chasing off all other males in extremely aggressive displays.

In Africa, 25 to 30 million years ago, the line of the Great Apes diverged from the Old World monkeys. Great Apes were the ancestral species of chimpanzees, bonobos, gorillas, orangutans and humans.

Within primates are instinctual traits that humans retained or discarded. Figuring out the ones we retained can tell us a lot about what lies at the core of our own evolutionary wiring, which underlies many (if not all) of our actions and how we build our societies.

GORILLA 'WARFARE'

Humans split off from the evolutionary ancestors of gorillas about 10 to 12 million years ago. While gorillas may look threatening, most gorilla aggression comes in the form of intimidation and displays, though they can make a very good job of defending themselves if mere threat doesn't cover it. By and large, it is warfare by bravado.

Gorilla hierarchies typically have female gorillas stay with the same group for life, while male gorillas are driven out by the leader of the group, the silverback, when they come of age, to wander as bachelors until they can construct female-populated groups of their own or supplant an existing silverback of another group. Male competition led to a high degree of sexual dimorphism, an evolutionary process where notable sexual differences gradually appear between biological sexes, with male gorillas being significantly larger on average than the females. Male gorillas also tend to kill infants that they did not sire, in order to increase the chances that their own DNA will become dominant instead.

Female gorillas make sure to form relationships with males to gain protection from these predators, not to mention the protection of their young from infanticide. Females which are kith and kin tend to stick together in sisterhood, being very supportive of each other's interests and safety. Female gorillas which are not related tend to compete aggressively.

Male gorillas are more likely to be hostile to each other, even when they are related. Competition and hostility are more common. With one notable exception. When male gorillas have been booted by a silverback from a female-populated group, they sometimes band together rather than roaming around solo, and when in exile they are much friendlier to one another, even engaging in mutual grooming and friendly wrestling. Some gorillas even eschew the harem completely and engage in the occasional bout of gay sex.

OUR CLOSEST COUSINS

Chimpanzees are our closest surviving evolutionary cousins. We share 98.4 per cent of our DNA with them. We split off from chimps through a last common ancestor about 5 to 7 million years ago. Chimps are smaller than humans, about 100 to 120 centimetres in height. But they are typically much stronger and more aggressive. Chimps have a brain that is three times smaller than a human's. Nevertheless, we can see a lot of similar instincts and behaviours in them. Not to mention inventiveness, genius and group politics. Chimps eat plants and insects, and not infrequently they are sighted hunting colobus monkeys. Males go around in packs to gain access to this food and to protect their territory from other groups of chimps. It is likely that territoriality is a trait passed down from our last common ancestor with chimps, but that sort of behaviour is hardly unusual in animals. What is novel is the organised way chimps do this.

Unlike gorillas, it is entirely common practice for chimp groups to be composed of a collection of males with a leader and a corresponding group of females, which also arrange themselves in a hierarchy. The leader of a chimp group can be the strongest and most aggressive, but not always. The chief must also be the most manipulative and savvy at maintaining alliances to support his rule. A chimp Machiavelli. As a result, it is sometimes the case that the leader is not the biggest bruiser per se, but a leaner, weaker politician who has managed to convince others to do his bidding. Other males have been known to team up and launch a violent revolution that overthrows and replaces the leader. This looks decidedly more like human politics.

Females have their own firm pecking order: some dominate, others submit to other females. The female dominance hierarchy also extends to offspring. Aggression towards a high-ranking daughter, even while she is young and weak, is punished by the dominant mother and her allies. The daughter is thus protected until she reaches the point where she can start forming dominance alliances of her own. There is a whisper of the hereditary principle in this behaviour, where one can gain additional privileges in a hierarchy because of who your parents are.

Meanwhile, chimp male dominance is entirely dependent on acceptance by the hierarchy of females. If they don't like you, you cannot be the group's leader. If you are already the leader and females turn against you, they will help to overthrow you and put a new male in your place. Even this has echoes of the 'soft power' held by elite women (such as Roman empress Livia Drusilla) in human history before modernity.

If you are higher up in the hierarchy, you get priority access to mates and food. Chimpanzee hierarchies are noticeably complex relative to these of other primates, and a larger brain was required by evolution to cope with all the social interactions necessary to maintain alliances.

Like many primates, chimps use tools. They fashion sticks to fish termites out of the ground for food. They use rocks as hammers. They use leaves as sponges to soak up water. They use branches as levers. They even fashion banana-leaf umbrellas. These techniques are taught, passing from adult to child. This counts as a form of social learning. Even a form of culture. They do not, however, build on these inventions generation after generation. Otherwise, over 5 million years chimp 'termite fishing' would doubtless have reached an industrial scale.

Chimps have language. Most chimp communication is carried out by gesture, but they do have a limited range of vocalisations. The limit is imposed by the chimp's physiology restricting the range of noises they can make and also their brain capacity. In captivity, chimps have shown a remarkable aptitude for memorising a wide range of written symbols.

Chimpanzees can also be highly violent. Male chimpanzees band together, roam their territory and see if they can find a lone chimp to beat up. They set about kicking and hitting the lone chimp. It is common practice for chimps to start tearing off bits of flesh – particularly the ears, bits of the face and, most shockingly, the genitalia. Warfare among chimpanzees is not a thing. They don't have the numbers or the coordination. But they are perfectly happy to patrol their territory and brutalise strangers. Coordinated violence on out-group individuals certainly is a trait held in common with humans.

BONOBOS

Chimpanzees are male-led and quite aggressive, because sex is distributed according to hierarchy. In dramatic contrast is their close cousin (and ours) the bonobo. Approximately 2 million years ago, two groups of ancestral chimpanzees were separated into different environments by a growing Congo River. The chimps to the south (which became bonobos) evolved radically different habits. Bonobos live in a female-led hierarchy where sexual activity is rife. Males are often physically stronger, but on the very rare occasion where a male shows aggression towards a female, a sisterhood of bonobos gang up on him and put a stop to it. Sometimes they scare him off with hoots and shouting. Sometimes they break his fingers. Females can also be violent towards each other when enforcing the hierarchy. But overall violence is a lot less due to the profusion of sex.

Rare for most primates, bonobos can have sex 'face to face', engage in fellatio and cunnilingus and perform 'French kisses'. Bonobos are highly sexed and masturbate every few hours. When greeting each other, bonobos have a tendency to touch each other's engorged genitals in what is referred to as a 'bonobo handshake' in order to reduce initial tensions. Because sexual activity is more common in bonobo groups, there is less reason for male aggression in the first place. When two groups of bonobos meet in the woods, the males in the group might get a little tense at first, but then the females from the two groups cross over and start having sex with the strange males. The intergroup tension, which among chimps would lead to fighting, ends in an orgy among bonobos.

It is therefore perhaps unfortunate that humans are more closely related to chimpanzees than bonobos. But while aggression, war and male competition are all present in humans, we seem to share a lot of sexual habits with bonobos and we even engage with 'make love, not war' on occasion (though relative to instances of 'making war', the more hippie periods of human history are considerably rarer).

But what is open to question is how many of the familiar chimpanzee traits humans inherited from their last common ancestor and which ones were culturally invented by humans much later.

If the more negative aspects of human societies are founded on evolutionary wiring, they might never be expunged. If they are culturally derived, then they can be unlearned in the space of a generation or two. So one must also ask: how did we continue to evolve in the 5 million years that separate us from chimps?

BIPEDALISM

5 million years ago, our ancestors still dwelt in African forests. Our last common ancestor with chimpanzees walked on the ground with bowed legs, using its arms on the ground for balance. It was better suited for quickly climbing trees to avoid predators (of which there were many in Africa) than for covering long distances on flat ground.

4 million years ago, the climate entered one of its dry phases. The forests shrank back, leaving woodlands and eventually the wide, open savannah of East Africa. Primates that ventured away from the forests to find food could no longer scamper up trees for safety and they had to range farther and farther to find food. As a result, our ancestors started to walk upright on two legs, evolving bipedalism.

Some of our first bipedal ancestors, the *Australopithecines*, were short, standing only about a metre tall, or just above three and a half feet. They looked a lot like chimpanzees, only bipedal. *Australopithecines* were largely herbivores, with teeth adapted for grinding tough fruits, leaves and other plants (which humans have inherited despite our later adoption of meat-eating). They may have occasionally scavenged meat from corpses, but they weren't really equipped to consume raw meat and did not yet have control of fire to cook it.

Because *Australopithecines* were bipedal, this freed up their hands for regular use of an even wider range of gestures, widening their range of language. Most communication happened by gestures and facial expressions, rather than vocally via grunts and yelps. Even today, many anthropologists and psychologists assert that the vast majority of human communication still happens via subtle gestures communicating sophisticated emotions and mental states, rather than by words. Free hands also allowed *Australopithecines* to carry tools and transport them from place to place. Enhanced language and more regular use of tools put evolutionary pressure on *Australopithecines* to increase their brain capacity to keep up.

'Lucy', one of the oldest known human ancestors

HOMO HABILIS

By 2.5 million years ago, *Homo habilis* evolved. *Homo habilis* did not stand much taller than *Australopithecines*, and their brains were only slightly bigger. But there seems to have been an increase in intelligence and inventiveness. *Homo habilis* were known to have hit flakes off stones to use them for cutting. And making stone flakes is difficult. Human archaeologists have tried to re-enact this activity and it is tricky, requiring a lot of trial and error. It requires some pretty beefy intellect, intentionality and the patience of a craftsperson. But there were limits. As important a breakthrough as stone-working was, we see very little sign of technological improvement over the million years that *Homo habilis* existed. We see invention. But we don't see the accumulation of invention generation after generation to make those cutting implements better or more diversified.

As for the social complexity of *Homo habilis*, it was arguably similar to that of *Australopithecines* or chimpanzees. Their groups remained quite small. But 2 million years ago, population growth caused *Homo habilis* groups to run into other groups more frequently. This put pressure on the brain to manage more frequent and complex social interaction, including alliance building, so that violence would not break out every time groups met. Strategies included gift-giving and intergroup marriage. The latter was particularly effective since it prompted two groups to have an interest in the continuation of a combined line of DNA. It is around 2 million years ago that evolutionary anthropologists estimate that monogamy (long held by New World monkeys) began to evolve in our own family tree in Africa. The fact that *Homo sapiens* have both successful and unsuccessful attempts at monogamy, in addition to polygamy and promiscuity, is testament to two strands of evolutionary wiring being in conflict with each other.

Another way primates bonded with each other and formed alliances was by grooming: picking the bugs and dirt out of someone's hair. We see this in our last common ancestors with Old World monkeys, going back 40 million years. But as group numbers grew, we couldn't groom everyone. There wasn't enough time in the day. So we began to 'gossip' or small-talk.

Homo habilis still had a very restricted range of sounds to form speech. But gesture allowed some communication, added to which they used pleasing sounds such as hums, grunts and yells to convey displeasure. There was an evolutionary advantage to socialising, putting selective pressure on these communication abilities to grow.

This may have been reinforced by sexual selection. Females may have chosen males who were able to express themselves in ways that either charmed them or convinced groups to follow them. Since the last common ancestor with chimps 5 million years ago, mating preference had been given to those males who were able to form alliances and ranked high in the group.

The pressure on enhanced communication to deal with growing social complexity had a profound impact on brain growth, which manifested in our next major ancestral species.

HOMO ERECTUS

1.9 million years ago *Homo ergaster-erectus* evolved. There is debate over where both *ergaster* and *erectus* should be classified in a single species that looks pretty similar. *Homo ergaster* usually refers to the earliest versions of the species that existed in Africa, while *Homo erectus* refers to the species as it travelled across the Old World. For simplicity, I will simply refer to both as *Homo erectus*. But this is not a position on the current debate about the taxonomic classification.

Homo erectus was taller than *Homo habilis*. It had perfected the art of bipedal locomotion. *Homo erectus* was definitely more comfortable crossing long distances than *Homo habilis*. In fact, *Homo erectus* would present a challenge to human bipeds today in stamina and running speed. Their facial structure looked far more human; if you saw one wearing clothes on a bus you might be forgiven for not really noticing anything unusual about them. Their body hair had receded significantly from that of earlier primates, leaving melanin of the skin protecting them from the African sun's harsh rays. In fact, in most major phenotypical aspects, *Homo erectus* was exceedingly human.

There is some evidence that *Homo erectus* lived in larger social groups than earlier ancestral species, and encountered other groups more frequently. There is also evidence that they had

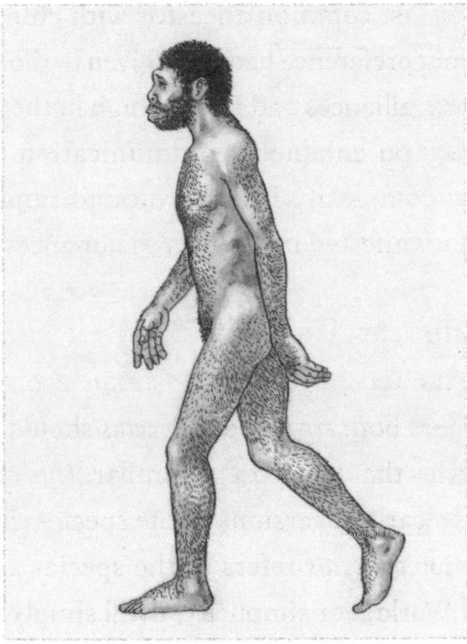

Homo erectus

controlled use of fire and were able to cook and eat meat. The consumption of meat was crucial to further brain development, since it packed more energy in a single morsel than a comparatively larger amount of vegetation. The most notable aspect of *Homo erectus* is that they had a decidedly bigger brain, roughly twice that of *Homo habilis* and 70 per cent that of modern humans.

A population boom led *Homo erectus* out of Africa and across southern and eastern Asia. They adapted to deserts, forests, and coastal and mountain regions. A species this adaptable certainly had to be advanced in intelligence. They became the first pan–Old World human species and they continued to exist for hundreds of thousands of years.

THE FIRST COLLECTIVE LEARNING?

There was very little technological improvement in the toolkit of *Homo erectus* in the first millennia after it evolved 1.9 million years ago. Then, 1.78 million years ago, *Homo erectus* invented a new kind of tear-drop axe in East Africa. This could just have been a one-off. For thousands of years *Homo erectus* did not tinker with or improve this tool. That was in keeping with every primate tool-user before them. Chimps, *Australopithecines* and *Homo habilis* were all bright enough to come up with new tools and to pass the techniques on to their offspring, but not to improve them generationally.

Yet with *Homo erectus* 1.5 million years ago in East Africa, we see the first glimmer of evidence of a revolutionary new ability. *Homo erectus* began to improve their hand-axes in quality, and converted them into multipurpose picks, cleavers and other kinds of implements.

This is hugely important in our story. It was the first sign of tinkering, accumulation of innovation, and improvement of technology generation after generation. Something known as **collective learning**.

Why does this matter? If there is a limit to how much one can invent, a species more or less stays the same for thousands of years until biological evolution changes them. Even with tool use, they are still stuck in the slow process of natural selection in order to raise complexity. However, if a species like *Homo erectus* could improve on existing technologies by tinkering with them – and with no major genetic change or evolution – and also spread out from their traditional habitats across the world, this is the sign of something new. It means this species was no longer dependent on biological evolution or the cruel Darwinian world to increase its complexity.

We have made the first tentative steps into the 'cultural realm', where the complexity-generating process of collective learning blazes along at a faster speed than biological evolution. Like express highways built atop older winding roads.

And collective learning had only just begun to evolve. A trickle would soon become a flood.

PART THREE

THE CULTURAL PHASE

315,000 years ago to present

7

HUMAN FORAGERS

Wherein Homo sapiens evolve from a long lineage • Collective learning is more powerful than ever before • Roughly 25 billion of us live in foraging communities spread over 98 per cent of human history • A genetic bottleneck reduces our gene pool to less than 10,000 individuals • Shortly thereafter we migrate across the world.

ACCUMULATION. THAT SINGLE WORD, more than any other, sums up what makes *Homo sapiens* different. The ability to accumulate more information with each generation than is lost by the next, also known as **collective learning**. Humans didn't get to where we are today because we are all super-geniuses. A cursory glance at politicians, celebrities or perhaps your in-laws is proof enough of that. A human raised alone in the wild would not have any remarkable advantage over other animals. Also, humans can only invent so much stuff in a single lifetime – when they're not too busy trying to survive, that is, which most people *were* throughout the majority of history.

Yet with generation after generation of invention, humans became something novel and unique in the biosphere. Like building blocks laid one upon the other. Slowly but surely, inventions stacked up, leading to dramatic changes in complexity within a few thousand years. In the blink of an eye in evolutionary time, humans have gone from stone tools to skyscrapers. That is the power of collective learning.

Isaac Newton said he stood on the shoulders of giants when it came to his own work on gravity (although, arguably, this was just rhetoric to cover up plagiarism). In reality, the 'giants' are actually composed of thousands and millions of inventors throughout human history. That is why the ability to accumulate innovation is what makes humans so unique. More so than just our raw brainpower or capacity for language and abstract thought. We have an unparalleled talent for remembering details of the past. Remembering our history.

FROM *HOMO ERECTUS* TO *HOMO SAPIENS*

Homo erectus showed the first sign of collective learning 1.5 million years ago. It was a very humble start. *Homo erectus* took tens of thousands of years to make moderate improvements to their stone axes. Nevertheless, collective learning had appeared in our evolutionary repertoire. Provided natural selection considered collective learning useful to survive, the skill would only grow more powerful with each new species.

Homo antecessor evolved 1.2 million years ago and moved into Europe in large numbers, requiring innovation to deal with the frigid and foreign environments there. They were about the same size in terms of height and body weight as *Homo sapiens*, but they had slightly smaller brains and a much more limited form of language.

Homo heidelbergensis evolved about 700,000 years ago in Africa and slowly spread over Europe and West Asia. They had even larger brains, which could fall roughly on the low end of the human average. It is likely they had a fairly acute sense of being able to distinguish sounds in speech, like modern humans, and a pretty intricate form of communication.

The Neanderthals appeared approximately 400,000 years ago and had brain sizes that rivalled that of modern humans, though they appear to have had limited capacity for abstract thought (contemplating and communicating things that aren't actually there).

Homo antecessor, Homo heidelbergensis and the Neanderthals all show clear signs of collective learning. They presided over the first systematised and regular use of fire in hearths, the first blade tools, the earliest wooden spears and the earliest use of composite tools where stone was fastened to wood. *Homo heidelbergensis* became the first hominine to colonise all of Eurasia. Neanderthals even adapted to climes that made clothing and other cultural innovations necessary for insulation and warmth. They manufactured complex tools, with prepared stone cores, producing a variety of implements – sharp points, scrapers, hand-axes, wood handles – with deliberate use of high-quality stone materials, and countless variations and improvements over time. All this invention and expansion across the Old World is a clear sign of collective learning growing evolutionarily stronger.

Then, 315,000 years ago, people that were anatomically identical to *Homo sapiens* first appeared in Africa. Why did *Homo antecessor, Homo heidelbergensis* and the Neanderthals all go extinct while *Homo sapiens* did not? Simply because *Homo sapiens* was the most talented at collective learning. For example, once *Homo sapiens* entered regions already populated by the Neanderthals, we outcompeted them for resources, likely killed many of them and notably also interbred with them (outside of Africa, a substantial amount of our DNA today includes Neanderthal genes).

Homo sapiens had the most advanced collective learning, the most diverse toolkits and were the most adaptable to new environments. We had big brains, a greater capacity for language and we were more capable of abstract thought, as evidenced by the fact that we alone painted cave art, used body paints, played music, wore decorative jewellery and demonstrated symbolic thinking. All of these traits complemented our capacity for collective learning as we built up huge knowledge bases about how to forage and survive in the hostile environments of the Paleolithic Earth.

Collective learning has two main drivers which make it stronger the more they are enhanced:

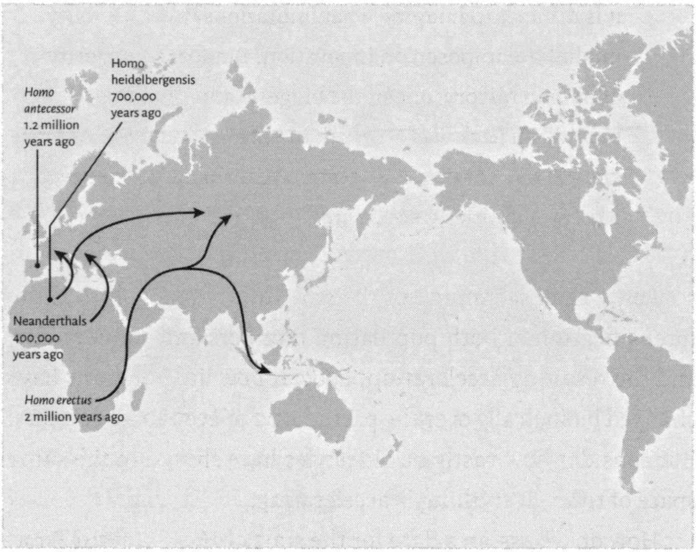
Homo heidelbergensis
700,000 years ago
Homo antecessor
1.2 million years ago
Neanderthals
400,000 years ago
Homo erectus
2 million years ago

1. **population numbers**: The number of potential innovators in a population. Not all of them will conjure an improvement in technology, doctrine or philosophy in their lifetimes. But the more people you have, the more rolls of the dice there are to increase the probability that one of those people will come up with either a minor or major innovation.

2. **connectivity**: In order to build on the ideas of the past, humans need access to them. This means either access to repositories of oral or written knowledge, or communication with other humans who possess that knowledge. Perhaps even collaboration with them. While today, with instantaneous communication via the internet and the equivalent wealth of knowledge of the Great Library of Alexandria being available on your phone, it is difficult to imagine what limitations in connectivity might have imposed on innovation. But for the majority of human history, one of the biggest things holding innovation back was access to the pool of wider human knowledge. For the first 300,000 years of our existence, our communities were limited to a few dozen foragers.

As we shall see, much of the story of human history is the intensification of both population numbers and connectivity, and the resulting acceleration. Look at how little humans have changed biologically over the past 10,000 or even 100,000 years. But consider how vastly our lifestyles have shifted in the same space of time. Everything is accelerating.

How do we assign a date for the start of *Homo sapiens*? First, there are the Omo remains, discovered in East Africa between

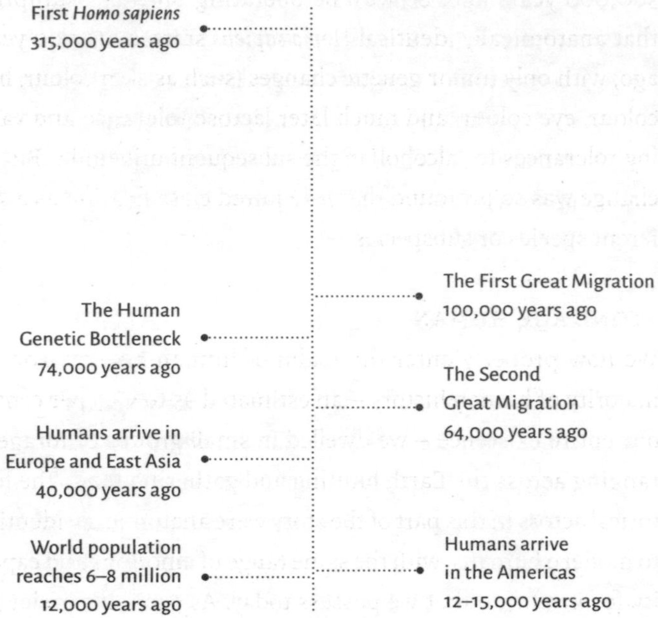
First *Homo sapiens*
315,000 years ago
The First Great Migration
100,000 years ago
The Human
Genetic Bottleneck
74,000 years ago
The Second
Great Migration
64,000 years ago
Humans arrive in
Europe and East Asia
40,000 years ago
World population
reaches 6–8 million
12,000 years ago
Humans arrive
in the Americas
12–15,000 years ago

1967 and 1974. Radiometric dating puts the oldest of these remains of anatomically identical *Homo sapiens* at 195,000 to 200,000 years old. Then, in 2017, we discovered more *Homo sapiens* remains in Morocco, dated to approximately 315,000 ago or older. So currently, 315,000 years old is the best supported start date for our species. But this number could grow larger in the coming years if new discoveries are made.

It is unlikely *Homo sapiens* underwent a rapid genetic change subsequent to 315,000 years ago that significantly enhanced its intellect or capacity for collective learning. They were using decorative beads in Africa prior to the Second Great Migration 64,000 years ago, mining for new materials 100,000 years ago, fishing 120,000 years ago, and using body paints around

300,000 years ago. So we'll be operating on the assumption that anatomically identical *Homo sapiens* started 315,000 years ago, with only minor genetic changes (such as skin colour, hair colour, eye colour, and much later lactose tolerance and varying tolerances for alcohol) in the subsequent millennia. But no change was so profound that it required classification as a different species or subspecies.

STONE AGE HUMANS

We now properly enter the realm of human history. For the majority of human history – an estimated 95 to 98.5 per cent of our entire existence – we dwelled in small groups of foragers, ranging across the Earth hunting and gathering food. The historical actors in this part of the story were anatomically identical to modern humans, with the same range of emotions and capacity for invention that we possess today. As such, it's easier for us to feel empathy with those in this part of the story, than the living beings of earlier periods. These people were *us*. If we had been born in that era, we would have behaved the same way. But human foragers lived very different lifestyles, in a world hit with ice ages and populated by a terrifying range of megafauna, from sabre-toothed tigers to carnivorous 3-metre-tall kangaroos. The past is indeed another country, if not another planet.

Taking the 315,000-year date as our starting point, approximately 100 billion humans have lived and died on the surface of the Earth since the origin of our species. Of those, 16 to 20 billion have lived since the start of the Industrial Revolution 250 years ago, with nearly 8 billion alive today, at the time of writing. An additional 55 billion are estimated to have lived between the beginning of agriculture 12,000 years ago and the Industrial Revolution. That's 71 to 75 billion people out of 100 billion.

That leaves approximately 25 to 29 billion people who lived between 12,000 and 315,000 years ago – the foraging era. For the majority of that time, most humans lived in Africa, and only in the past 64,000 to 100,000 years have a substantial number of humans lived in the rest of the world. We know that the Earth can only support about 6 to 8 million foragers living at any given time. For the overwhelming majority of the Paleolithic, our numbers were a lot fewer than 500,000.

Biologically and instinctually speaking, humans are best suited to the foraging lifestyle. It is what we were wired for. All the vast changes of the past 12,000 years since the invention of agriculture have not left enough time for us to evolve and catch up.

In short, we are cavemen in fancy shoes.

THE ICE AGES

The past 2.5 million years have seen numerous waves of cooling and heating, with long glacial periods (ice ages) interspersed with so-called interglacial periods (such as the one we are in now). There have been two or three ice ages since *Homo sapiens* evolved in Africa 315,000 years ago. During an ice age, large parts of North America, Europe and Asia are covered ice sheets, the average global temperature drops, previously lush climates in other regions of the world such as Africa dry out, and the sea level drops.

The second-to-last ice age started 195,000 years ago, when *Homo sapiens* were alive and well in Africa. It continued for a further 60,000 years, until an interglacial period began 135,000 years ago. This interglacial lasted for just 20,000 years, until approximately 115,000 years ago (interglacial periods are generally shorter than glacial ones). The 'Last Glacial Period' began 115,000 years ago and it was a particularly long one, lasting just over 100,000 years. It was in this world that humans migrated out of Africa and across the Earth.

At the height of the last ice age, about 30 per cent of the Earth's terrestrial surface was covered with ice. Where ice sheets did not spread, colder temperatures made forests turn to woodland or even deserts. Winters lasted longer than they currently do. While most humans 115,000 years ago lived in Africa, the conditions there were much more frigid than those experienced in Africa today.

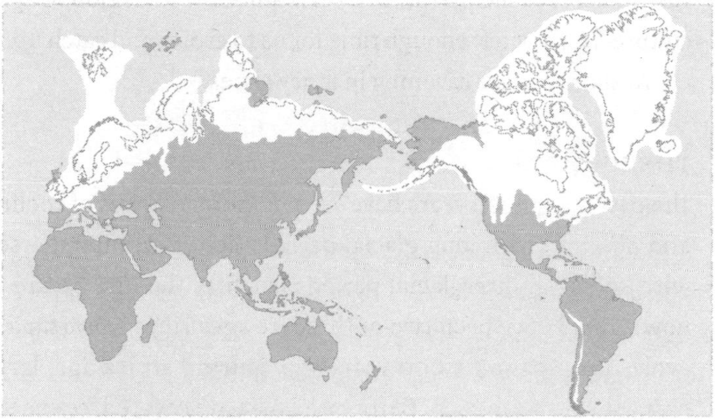

The world at the Last Glacial Maximum

THE GENETIC BOTTLENECK

Homo sapiens' foraging method of finding food remained the same for thousands of years: roam a territory hunting and gathering until the flora and fauna of the region become depleted, then move onto another region while the previous one naturally replenishes itself. Using this method, the entire surface of the Earth can support only 6 to 8 million foragers.

As human populations in Africa grew, they needed to find more food to support them. The answer to this problem was not to intensify the amount of food produced in Africa but to 'extensify' – by travelling further and further afield.

This may have prompted the First Great Migration of humans out of Africa 100,000 years ago into the Middle East, with some indications that they reached as far as India. These regions were still out of reach of the ice sheets. Despite this migration, the vast majority of humans remained in Africa.

We have indications in our DNA histories that human genetic diversity shrunk dramatically before the Second Great Migration. One possible explanation is the supervolcanic eruption that occurred during this period at Mount Toba 74,000 years ago. In the middle of the island of Sumatra, in present-day Indonesia, was a volcano. Where this volcano once stood there is now a lake. Or rather, a crater.

Mount Toba exploded with the force of 1.5 million Hiroshima-sized bombs and with the force of the nuclear arsenals of every single country in the world today. Multiplied by at least three. The eruption threw an unprecedented amount of rock into the atmosphere, scattering rubble and magma far and wide on a continental scale. A layer of volcanic ash, an average of 15 centimetres thick, settled over everything in South and East Asia, but also in India, Arabia and as far as East Africa. Much more ash was flung up into the atmosphere, darkening the skies and obstructing sunlight in an era already beset by an Ice Age. What followed may have been a decade of perpetual winter for the entire globe. It may well have reduced the population to 10,000 people. Or even as few as 3000.

In the past decade, the Toba hypothesis has been disputed by several scientists. I am still waiting for an alternative explanation for the bottleneck that is evident in our DNA. Hold on for the second edition!

Regardless, the genetic bottleneck tells us something very important about race. In a nutshell, humans today come from *at most* 10,000 people only a few tens of thousands of years ago. That is not enough time for significant genetic differences to occur between ethnicities. In fact, compared to other primates, modern humans today have an extremely low amount of genetic diversity. There is more genetic diversity between two groups of chimps separated by a few hundred miles than in the entire human race. We are not exactly an inbred species, but we are all pretty damn closely related.

THE SECOND GREAT MIGRATION

By 64,000 years ago, we headed out of Africa for a second time. Humans spread from Africa through the Middle East, down into India and Indochina within just a few thousand years. By approximately 60,000 years ago, humans had figured out how to use the land bridge that existed in Indonesia at the time (due to lower sea levels from the Ice Age) to make their way by foot and by rafting into Australia.

Stone-age seafaring is no easy task. The entry of humans into Australia is the foraging equivalent of landing on the Moon. Humans gradually spread across Australia over the next 20,000 years, heading into Tasmania by 40,000 years ago via another land bridge.

Also around 40,000 years ago, humans went north into colder climates, crossed the Caucasus Mountains and entered Russia, quickly making their way into Europe from the east. Most impressively, humans continued to spread into increasingly colder climates, and were in ice-age Siberia by at least

20,000 years ago. Consider the survival skills necessary in such an environment.

The entry of humans into the Americas requires a more searching statement. We are less sure exactly of how people got there. It seems clear that humans would have crossed the Bering Strait (which was then yet another land bridge) between Siberia and Alaska between 20,000 and 15,000 years ago, perhaps following herds of animals they hunted. But during the ice age there were huge ice sheets that would have prevented humans from travelling beyond Alaska. Then between 15,000 and 12,000 years ago, as ice sheets retreated, a passage may have opened up for foragers to travel through, moving south through the Americas. Another hypothesis is that humans may have bypassed the ice sheets by slowly rafting down the Pacific Coast. Or it may have been a mixture of the two. Either way, *Homo sapiens* became the first and only species of our genus to inhabit the Americas.

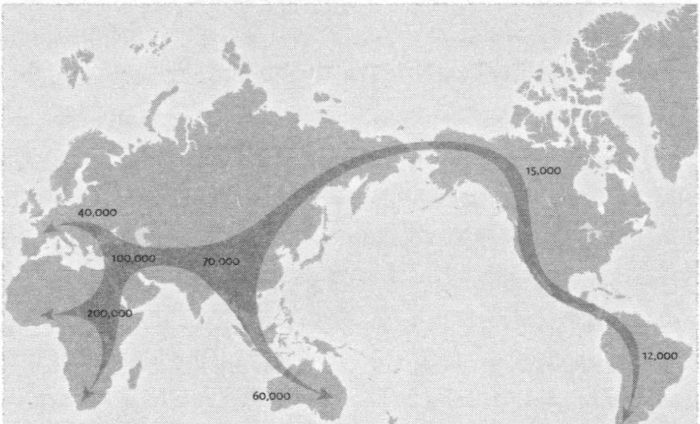
40,000
100,000
70,000
200,000
60,000
15,000
12,000

Human migration 100,000 to 12,000 years ago

OLD-FASHIONED WIRING

Humans were extremely well attuned to life as foragers, having existed in that state for 315,000 years and having evolved from previous hominines that also foraged for their existence. But we had to cope with potentially damaging situations that would end our genetic line there and then. Our instincts evolved accordingly.

In many ways, human instincts are designed to exist in small foraging communities. For example, take social anxiety. In modernity, there isn't really any good reason to have anxiety prior to giving a talk to a crowd of strangers or being nervous before a first date. There are millions of people out there in your city. You can humiliate yourself in front of hundreds of audience members or potential mates, and still go out the next day and try again with a different group.

This was not the case for foragers in the Paleolithic. A group of humans might be a few dozen in number, and you would spend your entire life with them. If you made a fool of yourself in front of a large crowd of people, you might be socially ostracised, which would reduce your access to food and mates, or you might be thrown out of the group altogether if they disliked you enough. If you made a fool of yourself in front of a prospective mate, they might tell all the others, and your DNA would be rudely ejected from the gene pool forever. In fact, this sort of danger within a small, close-knit social hierarchy goes back at least as far as our last common ancestor with chimpanzees, 5 million years ago.

In that context, it makes sound evolutionary sense to have an instinct that agitates humans in social situations. Much of our instincts have evolved in a similar way. And many of them are ill-suited to modernity.

FORAGING LIFE AND SOCIETY

Foraging means hunting and gathering. By and large, due to sexual dimorphism (differences in average body size and strength), male humans hunted and female humans gathered. But we know from studying modern foraging groups over the past two centuries that there would have been overlap between the two groups. Some females possessed the athleticism and specialised knowledge to be hunters; some males possessed the required knowledge of plants to be gatherers and/or were too old or infirm to hunt. Further, an individual might have been personally predisposed to one activity or another, just as today different people prefer different jobs. But exceptions do not make the average, and generally women gathered and men hunted. This general pattern goes back at least 2 million years.

On average, 60 per cent of food for foraging groups came from gathering. This was due to the feast-or-famine nature of hunting: you could go days without meat, then have a bonanza of several carcasses at once. Some social theorists have interpreted this percentage as indicating that females had equal 'hard power' in foraging groups, if not full egalitarianism in both political power and roles. But this is contradicted by modern studies of foragers. It also ignores sexual dimorphism and the fact that men are exceedingly more violent both towards women and each other. In short, it is no good bringing in a bunch of nuts and berries if a man can bash your head in with a rock. Human power hierarchies were/are not based on production alone – if they were, the farming peasant class would have ruled in the Middle Ages – but instead, on coercion, tradition and group loyalties.

But foraging societies also defy simple dichotomy. A low-ranking female was valued higher than a low-ranking male. If a female assaulted a male, there might be very little consequence; if a male simply insulted a female, he could be killed by other males in a foraging group. The only exceptions were the male leaders of a group, who could usually get away with more because of their rank. Power and rank mattered more than gender.

By and large, many foragers maintained monogamous relationships (particularly in ritualised marriages), but a few high-ranking men might practise polygyny (having more than one wife) due to their social status, and usually with religious justification. Beyond that, sexual and romantic relationships were just as turbulent and irrational in foraging communities as they are today. Emotionally, these humans were similar to modern ones, and thus probably experienced the same degree of emotions, from strong infatuations to tumultuous break-ups, jealousies and infidelities, all of which likely contributed to interpersonal violence.

Violence was more common in the foraging world than in any subsequent period. Studies of Paleolithic skeletons that show signs of deliberately inflicted violence as a cause of death indicate a 'murder rate' of approximately 10 per cent. Most of these corpses were male. This is much higher than the murder rate in any modernised country – or any society in the last 5000 years.

The typical foraging tribe went extinct every 200 years, on average – either due to genocide or through one cultural group of humans being conquered or absorbed by another group. A single human culture did not occupy a tract of land for thousands and thousands of years. Instead, for the majority of human history, biological or at the very least cultural genocide was the rule, not the exception.

When injury or sickness occurred, it was frequently a death sentence. And when foragers entered a region where food was scarce, they were at risk of starving. Simple things such as a broken bone, infected wound or rotten tooth could kill you. Infant mortality was high, with 50 per cent of children dying before the age of five. Furthermore, the fact that foragers had to be constantly on the move to find enough food for everyone meant that infanticide rates were quite high: around 25 per cent.

On the upside, collecting food only took up part of a forager's day. The average work day was 6.5 hours for foragers, compared to 9.5 hours on average for agrarian peoples, and the standard 8 hours for modern cubicle jockeys. The surplus time was used in socialisation rituals of various kinds, including feasting by the campfire, dancing and the all-important politics of mating.

As a result of all the different sources of food, the diverse diets of humans (when times were good) actually left foragers quite healthy. The nomadic life also meant there was little opportunity for as many viruses and contagious diseases to grow, meaning foragers were much healthier than humans in the subsequent agrarian period. All told, one could reasonably make the argument that humans lived better in the foraging era than at any time prior to the days of developed countries in modernity.

By the time humans had spread across the world 12,000 years ago, we had multiplied into a population of 6 to 8 million people. While humans had already proven themselves a highly adaptive and powerful species, and the sophistication of their ideas and toolkits were unprecedented in the *Homo* genus, a great revolution was just around the corner. It would not only put humans on a course towards ancient and modern history but trigger a great acceleration in just 12,000 years – a blink of an eye in the timescales we have covered so far – and the acceleration has not stopped but only grown faster in modern times. It is worth bearing in mind that we stand on the precipice of further revolutions with potentially godlike and Universe-changing results. And it all started here, with a few million clever primates chipping tools out of stone.

8

THE DAWN OF AGRICULTURE

Wherein humans grab more energy flows from the Sun via photosynthesis of crops • Crops support more people over a smaller land area • More potential innovators living at close quarters accelerate collective learning • Complexity goes absolutely nuts.

AFTER SEVERAL TENS OF THOUSANDS OF YEARS of migration, humans had spread into every major world zone on Earth. Following the end of the last ice age, the world's foraging population reached its zenith at approximately 6 to 8 million people. The largest population was in Afro-Eurasia, with an estimated 5 million people, followed by the Americas, with an estimated 2 million people, and Australasia, with 500,000 to 1 million people. Humans had yet to settle in much of the Pacific islands until between 4000 and 800 years ago.

Due to the end of the last ice age 12,000 years ago, the Fertile Crescent in the Middle East experienced a greening and abundance of food so that generations of foragers did not have to migrate in order to find it. Such territories are called 'gardens of Eden' by historians and archaeologists. Foragers switched to semi-sedentary lives collecting surrounding vegetation and hunting animals, while reducing the amount of distance they travelled over the course of a generation.

Then, as populations boomed and food became scarcer, these foragers were forced into what archaelogists call the **trap of sedentism** and were compelled to domesticate plants and animals in order to prevent themselves from starving. This was the start of farming, where deliberate cultivation of food resources supported larger populations and a greater population density. The practice of agriculture spread into Egypt (or was independently conceived there), across the Middle East and gradually into Europe.

In China, similar 'gardens of Eden' emerged roughly 9500 or 10,000 years ago in the valleys of the Yellow River in the north and the Yangtze River in the south. The end result was also similar, with the inhabitants of East Asia beginning to cultivate plants and animals in order to support larger population numbers. Correspondingly, agriculture spread into Indochina and Japan. With agriculture now in the Middle East and East Asia, a combination of these practices gradually converged and met in South Asia – and the Indus Valley in particular.

The barriers of the Sahara Desert and the world's oceans prevented the practice from spreading into certain regions of the globe. In West Africa, a similar 'trap of sedentism' happened independently in the river valleys of the Niger and the Benue around 5000 years ago, subsequently spreading across West Africa. This region has the highest population densities in Africa to this day. Millennia later, the practice of agriculture was taken towards the southern tip of Africa, with mixed results, where many Africans insisted on maintaining traditional nomadic lifestyles right up to modernity.

Meanwhile, the trap of sedentism occurred in Mesoamerica 5000 years ago, and gradually spread south into Peru and north into the Pueblo societies of the south-west United States. Most intriguingly, an independent invention of agriculture also occurred in New Guinea 5000 years ago, with populations remaining quite small. In Australia, foraging remained the primary mode of life, with the notable exception of fire-stick 'farming' (the practice of burning down large tracts of forest to clear paths, kill wildlife and encourage fertility and rejuvenation), which was quite productive, and also instances of aquaculture in South Australia that sustained a few sedentary populations of several thousand people.

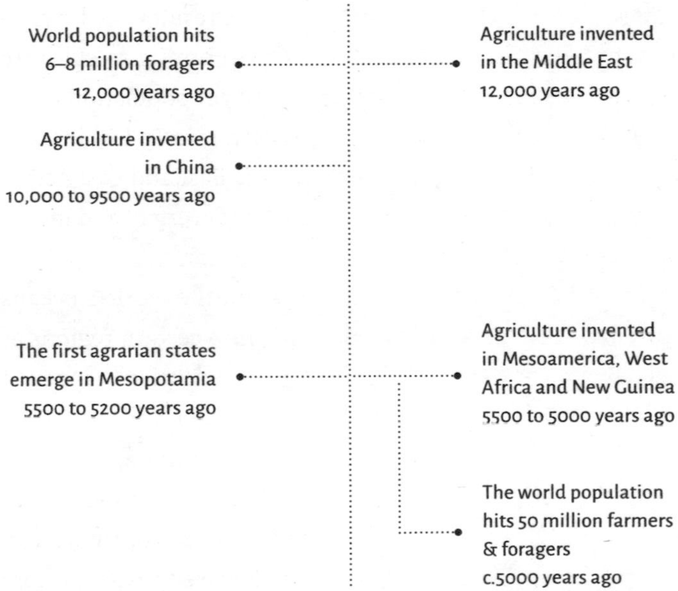
World population hits 6–8 million foragers 12,000 years ago
Agriculture invented in the Middle East 12,000 years ago
Agriculture invented in China 10,000 to 9500 years ago
The first agrarian states emerge in Mesopotamia 5500 to 5200 years ago
Agriculture invented in Mesoamerica, West Africa and New Guinea 5500 to 5000 years ago
The world population hits 50 million farmers & foragers c.5000 years ago

THE ASTONISHING COMPLEXITY OF MERELY PLANTING A SEED

By several different metrics, the beginning of agriculture marks a new threshold of complexity in our story. For starters, the amount of energy flows humanity could capture to sustain their own complexity more than doubled, from about 40,000 erg/g/s in a foraging community with controlled use of fire to an average of 100,000 erg/g/s for the average pre-modern farming community. Bear in mind that the Sun itself scores only 2 erg/g/s in energy flows, single-celled life 900 erg/g/s, and most multi-celled life achieves between 5000 and 20,000 erg/g/s, depending on what it is doing. Structurally, an agrarian society is not just a network of cells in a single organism but a fragile web composed of many different organisms: humans, plants and animals. And the societal web was one of the most structurally intricate and densest in energy flows in the entire Universe. If our history had stopped

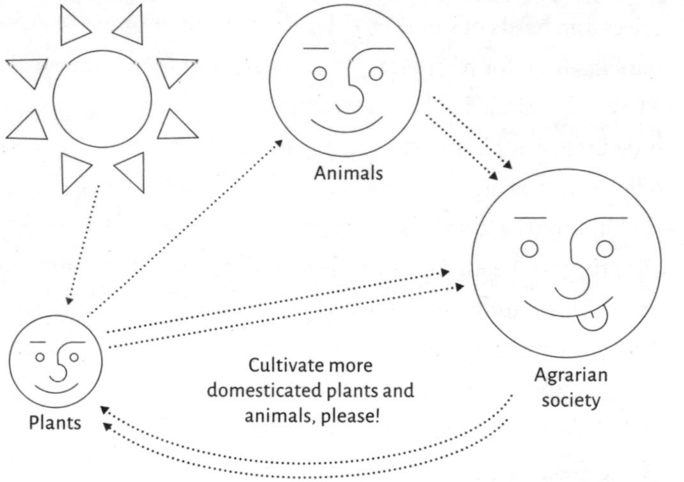
Animals
Agrarian society
Plants
Cultivate more domesticated plants and animals, please!

in the Neolithic 10,000 years ago, this would still be a notable milestone in the history of the Universe.

In short, on one tiny rock in our solar system, couched within one tiny pinprick of unequally distributed energy that emerged shortly after the Big Bang, that little dot of energy was progressively growing in intensity, becoming denser and denser, more complex than anything in the vast cosmos around it.

Both foragers and farmers capture the overwhelming majority of those energy flows from the Sun, as do most organisms on Earth. Foragers will wander across a region and pick vegetation (which gets its energy from photosynthesis) and kill animals (which also eat those plants) and cook plants and animals by burning wood (from trees which also drew on the Sun).

But farmers would not rely merely on what grew naturally in the uncultivated wilderness. Some of those things were not fit for human consumption but took up valuable space. So farmers cleared forests, nurtured soil, irrigated fields and planted rows and rows of high-energy consumable plants, with which they fed themselves and herds of hundreds of domesticated animals, which the humans used for their wool, milk and meat instead of picking off a few wild animals in hunting parties. Humans started selectively breeding both plants and animals to become even more energy-efficient: fatter animals for meat or grain plants with higher yields.

This marks a shift in nature where a species is no longer adapting to the environment but adapting the environment to them. Ultimately, this new way of life supported more people. Agriculture drastically increased the amount of people the land could support – by 1000 to 10,000 per cent per square kilometre relative to the foraging era. Suddenly the carrying capacity of the entire surface of the Earth wasn't 8 million foragers, but 80 million – and eventually 800 million – farmers.

The impact of increased energy flows and ultimately more people had a positive feedback loop on collective learning. With agriculture, there were more people (potential innovators), who increased the probability that a few of them in each generation would come up with innovations. Some of these innovations raised the carrying capacity of the population still further, whether through a new mode of farming, a new crop or a new tool or technology. As a result, there were more people, which led to more innovation, which kept the process accelerating.

Not only did innovations speed up in agrarian societies compared to foraging ones, but farming regions quickly became the most populous regions on Earth. Instead of dwelling in nomadic groups of a few dozen people, humans began to live on farms with populations that large and congregating in villages that had hundreds of people. Foragers not only found it difficult to keep pace with the technological evolution of agrarian societies,

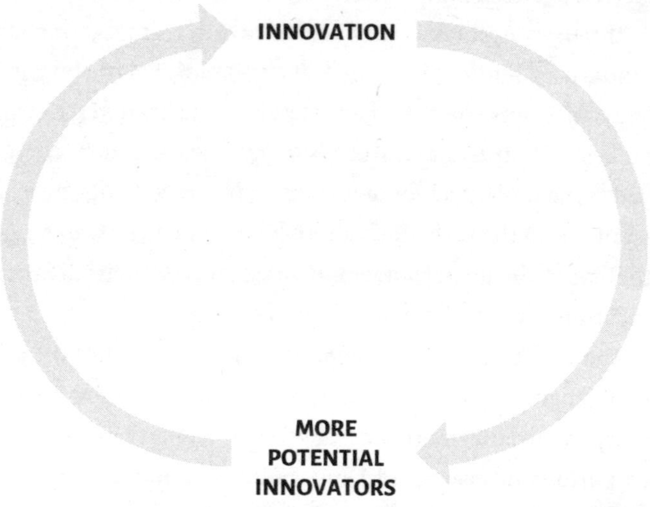
INNOVATION
MORE
POTENTIAL
INNOVATORS

but quickly found themselves outnumbered as well, as the land they had freely used to hunt and gather was gradually whittled away by sedentary farms. This forced many of them to move further away or take up farming themselves. As a result, foragers starved or enacted raids and violence on farming communities (with the danger of retaliation). For the next 12,000 years, wherever agrarian societies appeared, this tragedy played itself out on the borderlands with foragers.

BOOZE, DISEASES AND EXCREMENT

From approximately 12,000 to 5000 years ago, agrarian society (where it existed) was composed of farms and villages only. No cities, no states, no armies, no writing, no royal dynasties. None of the trappings of conventional history. A world of farms and villages prevailed for 7000 years. For such a long stretch of time, increasing numbers of humans tried their hand at farming, with all the maladies that accompany it. This era before states is referred to as the Early Agrarian Era.

The Early Agrarian Era is generally characterised by a poor standard of living compared to the Paleolithic or the Era of Agrarian States (though the latter certainly varies by circumstances). For the entirety of the Early Agrarian Era, farmers used stone tools. And while the implements were inventive and testament to the power of collective learning, they weren't very efficient. Nor did early farmers have very good fertilisers or irrigation.

As a result, the carrying capacities of the Early Agrarian Era were generally low. This meant that after the initial burst of plenty the first farmers may have enjoyed, there were numerous periods of overpopulation, malnutrition, starvation and famine. Animal power was not adequately harnessed (no pun intended) in this period, so most of the planting and tilling was conducted by humans with the aforementioned primitive stone tools. It was down to adult and child labour (another advantage of having more kids compared to foraging) to cut down forests with stone axes, break open rough soils with stone hoes and cut down the crops for harvest with hand-held scythes made of stone or bone.

Nor were the fertilising benefits of animal poo fully realised at the time, meaning that soil would rapidly lose nutrients and render cropland useless for several years. Early agriculture was highly dependent on natural water sources (rivers), since there was neither the technology nor the manpower to conduct sophisticated irrigating techniques to make wider stretches of land suitable for growing crops, meaning the amount of land that could be effectively farmed was limited.

Even without famine, the conditions of the Early Agrarian Era were fairly disgusting compared to the Paleolithic. Foragers had a reasonably diverse diet, there is every indication that in normal circumstances they cleaned themselves regularly, and because they lived constantly on the move in small communities without domestic animals, there were few communicable diseases. In the Early Agrarian Era, on the other hand, humans were sedentary and remained within the same few square kilometres for their entire lives. This meant that the waste from their food products (rotten vegetables, carrion and the offal of dead animals) and the improperly disposed of results of digestion (human and animal excrement) stuck around, often close to the home, creating unsanitary conditions from which people could fall ill. As a result, typhus and cholera became huge problems and were quite deadly and contagious. Typhus was caused by a virulent bacteria that could be transferred from person to person via mutual contact with food and could also spread through a water source. Once infected, a person was highly contagious and suffered from fatigue, swelling, pain, fever, delirium, hallucinations, heart problems, ulcers and intestinal bleeding. Cholera is caused by a bacteria that infests the lower intestine and causes extreme bouts of diarrhea and vomiting, so dehydrating the victim that the skin retracts, the eyes sink, the skin turns blue and the person eventually dies. Other viruses and poxes were whipped up in larger populations living in close contact with each other and with their animals. These spread by coughing and sneezing, and the resulting poxes disfigured the skin, and caused swelling in the brain, seizures, fevers and death.

It did not help that humans frequently bathed and defecated in their own water supplies, as did their neighbours and many of their domesticated animals. Bathing did not necessarily get one clean and one could fall ill from it, so in some regions personal hygiene fell of fashion, with regular bathing actually being considered unhealthy (in other areas, groups still bathed regularly by custom). Lack of bathing further exacerbated health problems. And there were no reliable soaps or antibacterial agents for thousands of years. People became accustomed to body odour and (as a result of diet and lack of dental hygiene) bad breath and rotten teeth.

There was also the problem of contaminated drinking water from these same causes, which made drinking water alone quite unhealthy. The happy result of this (or unhappy result, depending on your perspective) was the invention of alcohol. Via fermentation, watered-down meads, beers and wines were made safer to drink than pure water. That is not to say that humanity spent the next several thousand years of history drunk off their heads (though that would be an amusing explanation for some decisions): most beverages were not as potent as the ones that began to be distilled, commercialised and sold for recreational drug use purposes in the nineteenth and twentieth centuries. The average alcohol content of pre-modern beer was around 2 per cent. But the spectre of alcoholism in 10–25 per cent of the population actually goes further back than the start of human farming. It goes back 66 million years when our shrew-like ancestors would eat rotting fruits and wild grains, and consume very small amounts of fermented alcohol as a result, getting a small reward of dopamine in their tiny brains. The pleasure response was evolved to encourage this behaviour, prompting our ancestors to consume rotting stuff in order to stave off starvation and increase our chances of survival. Once we started mass-producing alcohol, we essentially glutted ourselves and this neurological response went into overdrive.

Early farmers also lived at close quarters with domesticated animals, sometimes even within the same dwelling, and the transference of viruses and bacteria between humans and their domesticated animals bred avian and swine flu that could rapidly sweep through and ravage a human population. Food and waste products also attracted pestilence. Rats, fleas and cockroaches became commonplace. The regular denizens of filth kindly shared a new range of diseases, including various forms of infection, dysentery and dreaded variants of the plague.

Sound appealing? If at any point in our story thus far you thought 'complexity' was synonymous with 'progress', then let the Early Agrarian Era disabuse you of that notion.

IT TAKES A VILLAGE ...

Putting the famines, pestilence and diseases that could make you shit yourself to death aside, these early farming societies were able to support way more people per square kilometre than the foraging cultures from which they emerged. The result was an acceleration of collective learning and a resulting rise of complexity.

In the foraging era, the centre of gravity for a society was the family. Kinship was the primary mode of maintaining governance, and alliances between groups were maintained by ritualised intermarriage. The rise of agriculture added another layer to this societal complexity. The farm still consisted of the family, with each member engaged in daily duties for subsistence, and with intermarriage occurring between families on neighbouring farms. But the social life of the agrarian society converged in the village, a place populated by a few hundred people who would gather to exchange things (agricultural goods, tools and information) and engage in the governance of affairs that affected the wider community (crop yields, problems arising from the weather, the possible threat of raiders and the resolution of disputes between families). Villages were also places where grains could be stockpiled in case the wider community was struck by famine. There even appears to have been the development of religion in the Early Agrarian Era, with villages partaking in increasingly elaborate burial traditions for their dead. These burials yield an array of jewellery and other decorative items, which may well have denoted status, and thus the increasing sophistication of hierarchies.

In terms of violence, the majority of it doubtless remained interpersonal, as in the foraging era. But with sedentism and the introduction of land claims, crop yields and possession of livestock came conflict over property. This may have manifested itself in either thievery by neighbours or land disputes between them that were arbitrated by the wider community.

There was also the new problem of raiders: neighbouring cultures (either other sedentary farmers or non-sedentary foragers) who would rampage across a farming region, taking crops, livestock and tools, and perhaps even kidnapping women and children. The earliest agrarian settlements, such as the village at Abu Hureya in Mesopotamia, which housed sedentary farmers 10,000 years ago (8000 BC), don't show much sign of defensive apparatuses. But as the agrarian era dragged on, farming communities begin to build walls, ditches and watch-towers around local villages. One of the most impressive examples of this is the village of Banpo in China, which lasted from 7000 to 5000 years ago (5000 to 3000 BC), and where all the dwellings were clustered in a group behind a wall which was surrounded by a ditch.

An excavation in Banpo Neolithic Village in Xi'an, China.

An even older example is the settlement of Jericho in the Fertile Crescent, which was converted into a farming village 11,500 years ago. The original settlement had no structural defences but had a cluster of houses built upon a freshwater spring that was directed via primitive irrigation ditches towards the surrounding 10 square kilometres of farmland. By 10,000 years ago, however, a wall was erected around the village.

The purpose in both of these cases seems clear. In a village where trade was conducted between farmers, where some grains may have been stockpiled, and thus there would occasionally have been a cluster of resources, defences were required in order to prevent large raiding parties from coming along and 'redistributing' the community's wealth. Please note, however, that these defensive structures in a village do not necessarily denote the existence of large-scale warfare – that was beyond the resources of Early Agrarian societies. Instead, these raiding parties were opportunistic, skirmishing with defensive militias formed by local farmers on the frontier.

POWER AND HIERARCHY

In order to organise this extra layer of society and cope with the many legal and defensive needs of a denser agrarian community, we see the microcosm of **entrenched hierarchy**, which is any hierarchy established on more than personal rule: that is, with a ruling class who you have probably never met. Bear in mind that during the Early Agrarian Era, the overwhelming majority of the population were engaged in subsistence farming to stay alive. A very small minority of people took on positions of authority to arbitrate disputes and organise infrastructure projects that could not be executed by a single individual or family.

The appointment of such authorities within a farming community would have come about in one of two ways (or both). The first, and most likely the earliest, was 'bottom-up power'. When we discuss power, we are talking about an individual or council of people possessing the authority to issue commands and have reasonable expectation that those commands would be carried out. If you want to translate this into more universal terms, it is the direction of energy flows in the form of food or human effort towards a certain goal laid out by an individual in authority.

In the 'bottom-up' scenario, the farming community would appoint an experienced or sensible individual, usually an elder or council of elders (in Latin, *maiores*, from which we get the term 'mayor') to arbitrate disputes and make decisions for the community. These decisions would have bearing on the entire community (the system of energy flows). And in order to afford the time to make these decisions and shoulder these duties, the elders might be given food they did not farm themselves, so they would spend less time on subsistence living. At first, these positions were appointed on basis of merit, and decisions were obeyed by the community without a great deal of coercion beyond the interpersonal and the social browbeating of an individual or minority faction that did not want to play ball.

In this sense, the hierarchies of Early Agrarian societies were not much different from foraging ones, or even most primate ones. All primates have a dominance hierarchy of some sort. The difference is that once an agrarian population extended into the hundreds or thousands, it was difficult for an elder or group of elders to maintain dominance merely by being the strongest or by maintaining the strongest interpersonal alliances. There are only so many people within an agrarian community with whom the reigning authority could form a personal relationship.

Instead, a power structure might involve formalised procedures of bestowing power, by vote, by inheritance or by religious ritual. And in order for the commands of such an authority to be obeyed, the elder might quickly require a voluntary or paid group of enforcers.

Which brings us to the second method of establishing power, the 'top-down' method. This is where consent from a community is not necessary, because the authority of an individual or council is backed by the threat of violence. By the time agrarian settlements began constructing defences, there would have been a militia or group of men capable of violence in numbers. These groups would be used not just against outsiders but also against members of the community who did not obey commands or accept the judgement of arbitration in a community dispute. These enforcers needed extra energy flows as payment for their efforts, so they too did not have to spend their entire time farming. In order to maintain this cycle of energy flows, the elder could always use the enforcers to collect further tribute from the surrounding population. All this would have happened gradually, with the veneer of legality and consent within the village.

Please also bear in mind that Early Agrarian societies did not have the long history of ideological conditioning that predisposes us towards democracy. For instance, inheritance of authority might have seemed to them the much more natural course of action. The transition from a bottom-up, democratic (or at the very least meritocratic) appointment of a reigning authority to a hereditary, entrenched, aristocratic hierarchy could actually happen quite quickly.

This is not such a departure from our instinctual primate past, where chimpanzees maintain alliances based on inheritance, with the offspring of high-ranking members inheriting the alliances and protection once afforded to their parents. As such, the time elapsed between 'bottom-up' and 'top-down' approach may not have been equally distributed, though the exact timing of the development of leadership traditions (democratic, meritocratic and hereditary) differed by region and culture.

APPROACHING 'CONVENTIONAL' HISTORY

As unpleasantly familiar as all these power machinations may appear to be, we must remember that most humans in this era lived in small, close-knit farming communities with fairly well-adapted values. Including close families and friendly neighbours. Just as small foraging communities by and large were stable and liveable for many of the humans living in them.

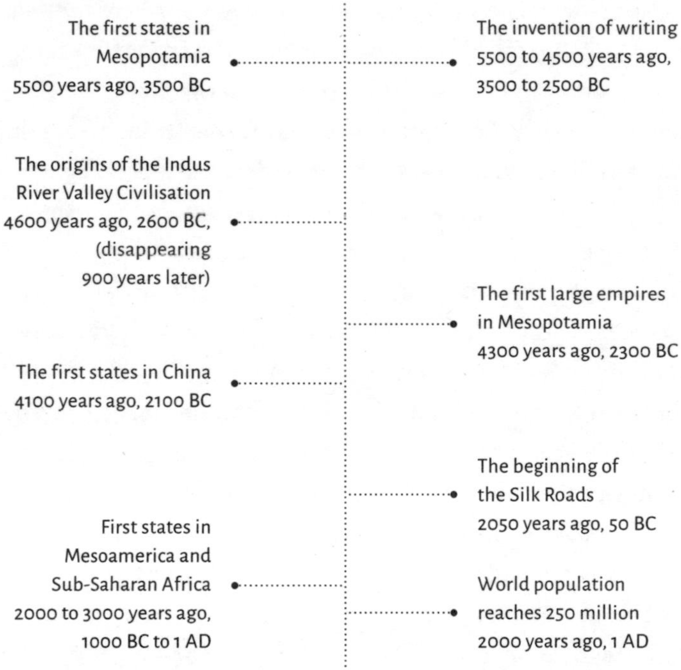
The first states in Mesopotamia 5500 years ago, 3500 BC
The invention of writing 5500 to 4500 years ago, 3500 to 2500 BC
The origins of the Indus River Valley Civilisation 4600 years ago, 2600 BC, (disappearing 900 years later)
The first large empires in Mesopotamia 4300 years ago, 2300 BC
The first states in China 4100 years ago, 2100 BC
The beginning of the Silk Roads 2050 years ago, 50 BC
First states in Mesoamerica and Sub-Saharan Africa 2000 to 3000 years ago, 1000 BC to 1 AD
World population reaches 250 million 2000 years ago, 1 AD

Just as your own communities today have the ingredients to produce healthy, happy lives despite the thunder and disgrace of wider politics. Life in any period, really, is what you make it.

The commonalities that unite all periods of human life were already in place at this point, because we have functionally been the same human animal across the millennia for over 315,000 years. The most notable aspect of the period of 'conventional' history of the past 5000 years is how rapid and 'unconventional' the pace of change has been during it, and how pronounced the rise of complexity would be from here on.

9

AGRARIAN STATES

Wherein the first agrarian states arise • The world population grows dramatically • Cycles of rise and decline blight human history • Trade between states enhances collective learning • The evolution of printing sends knowledge sharing into overdrive, circulating among larger segments of the population.

WE ARE NOW STARTING CONVENTIONAL HISTORY. Nine chapters in. More startlingly, we will be covering the majority of conventional history (the first 6000 years) in one chapter. It's possible to do this by following the broad overarching patterns of complexity and collective learning. These patterns serve as a kind of 'universal acid' for the jumble of names, dates and events that form the totality of human affairs. Just like Darwinian evolution helps us make sense of the grim slaughter of billions of species in the fossil record.

Agrarian states still harnessed unprecedented power from the Sun for their crops and livestock, with 80 to 90 per cent of people remaining farmers. But collective learning gradually made agriculture more efficient, leading to the spread of farming across the Earth and the appearance of something new:

cities, bureaucracies, armies, artisans, scribes and rulers who did not engage in farming. The next level in structural complexity. Collective learning caused the world's population to increase, but it did not keep pace with agrarian birth rates, leading to recurrent population crises which kicked off worsening civil violence and even the fall of empires. These population cycles which influenced political events are known as 'secular cycles'. Those trends form the deeper tides which propelled much of the 'swirling foam' atop the waves of conventional history.

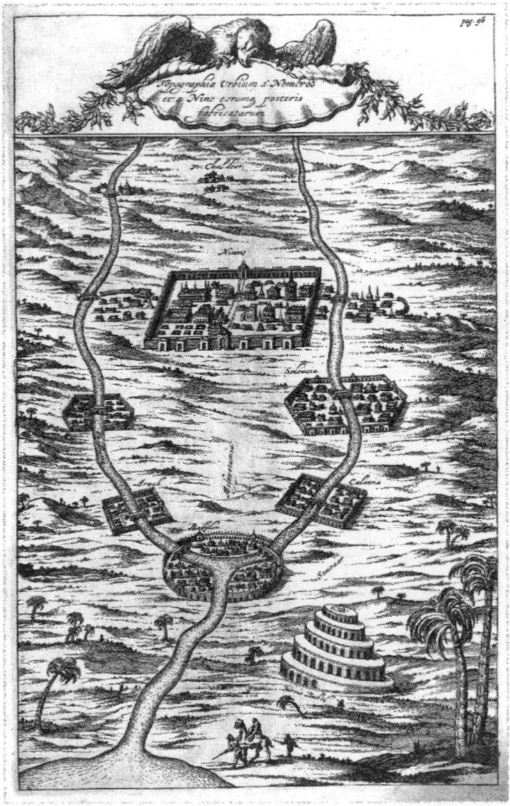
Topographia Urbium a Nimrodo
et a Nino eorumq posteris
fabricatarum

A seventeenth-century map of Babylon and Nineveh

RISE OF THE CITY

By 5500 years ago (3500 BC), the world population had grown from 8 million foragers at the invention of agriculture to 50 million people. That was a lot more potential innovators for collective learning, and the pace accelerated accordingly. The transition from the Early Agrarian Era to the Era of Agrarian States (starting 5500 years ago) is defined by:

1. the appearance of large cities with a **division of labour** (non-farmers being supported by surplus food crops)

2. the appearance of **writing**

3. the beginning of **secular cycles** (driving the rise and fall of empires).

In order to support a city, where many of the inhabitants are not engaged in farming, you need to grow surplus food in the countryside. Collective learning got to work in the Fertile Crescent around 7000 years ago. Hardier tools made of soft metal slowly replaced those of wood, stone and bone. Higher-yield crops were selectively bred by farmers over thousands of years. Irrigation introduced water to soils that were normally dry, unlocking previously untapped nutrients for plants. Using animals to plough fields cut up the soil much faster than any human could. Combined with favourable climates in the region 6000 years ago, farming productivity exploded by leaps and bounds. This surplus food allowed villages and towns to grow larger and larger.

In Sumer, the town of Eridu grew from a farming village to a city of 10,000 people by 5500 years ago (3500 BC). Between 5500 and 5200 years ago, a number of cities that size sprang up. But these were not as large as Uruk, to the north-west of Eridu. Fifteen times larger in land area, Uruk had up to 80,000 residents. This was a permanent human settlement on a scale never seen before.

With growing collective learning came more crops, which supported increased societal complexity. Uruk had a very pronounced division of labour, where non-farmers were fed by an increasingly efficient agrarian surplus. The city was ruled by a caste of priests headed by a priest-king. Under them were scribes who handled the complex logistics of the city. Palaces and temples were built by a large artisanal and labour force numbering in the thousands. Soldiers maintained law and order and manned the city walls. The city had a burgeoning linen and wool industry, many wealthy merchants, and slaves who were coerced to perform work as household servants or labourers. Outside the city, farmers would have constituted roughly 90 per cent of the population and the priests owned 30 to 65 per cent of the land. A good portion of farmers, too, would have been slaves.

Slavery appeared almost immediately with large human settlements. If there were enough crops to support a ruling class, and enough crops to feed soldiers to protect them, there were enough troops to coerce people to do work against their will. Often there were pretexts to legitimise slavery: someone got into debt, or was a criminal whose crimes were not heinous enough for execution, or was of a foreign religion or ethnicity. But in the majority of cases, slaves were enemy captives from a war. For over 5000 years, until just a few centuries ago, slavery was the rule for all agrarian states and abolition was the *very* rare exception.

Warfare began. Sumerian cities needed farmland to feed their populations and keep them wealthy. Armies of thousands of troops began to form for the first time in human history.

Uruk dominated in the period 5500 to 5000 years ago. Thereafter, increased competition from other city-states led to horrific violence. Uruk was conquered and sacked by its rival city Ur 4550 years ago (2550 BC). There had always been violence in primate dominance hierarchies, even in human foraging ones, but now the bloodshed occurred on the scale of thousands of dead or enslaved people, with no sign that the cycle would ever stop repeating itself.

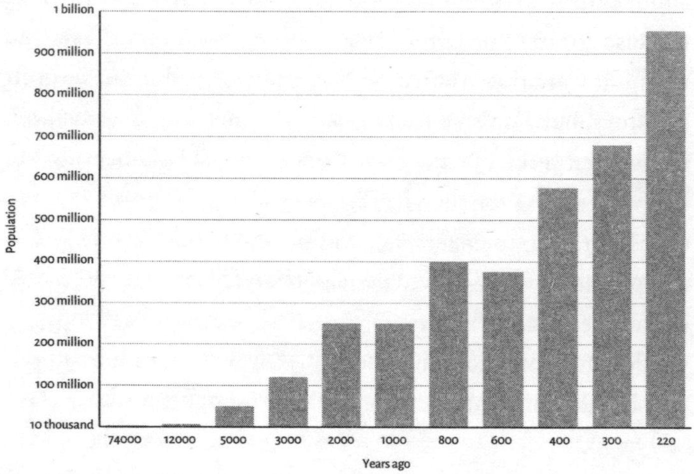
1 billion
900 million
800 million
700 million
600 million
500 million
400 million
300 million
200 million
100 million
10 thousand
Population
74000
12000
5000
3000
2000
1000
800
600
400
300
220
Years ago

Population growth from the genetic bottleneck to the Industrial Revolution

RISE OF WRITING

Uruk has the honour of providing us with the oldest surviving writing, inscribed by sticks into clay tablets dated to 5500 years ago (3500 BC). These writings discussed agricultural produce and livestock. From 5500 to 4500 years ago, Sumerian writing evolved from the pictographic (where symbols had no relation to how the word was spoken in Sumerian) to a wealth of syllabic symbols for complex songs, poems and histories, along with the addition of a system of numbers. Writing patterns would undergo similar evolutions in other agrarian states as they arose and evolved across the world.

In terms of collective learning, the advantages of written records are fairly self-evident. Instead of passing on all knowledge by oral tradition – where if there is a generation with whom it is not shared, it disappears – written records can slumber in an archive for centuries, only to be rediscovered. One can also communicate more complex and abstract information than if it were transmitted orally. This would include details of a history, but also calculations made in mathematics. All told, the written record made it less likely that the knowledge would be forgotten, as frequently happened in the foraging era. The only limitation to collective learning at the time was that very few people beyond scribes and priests would have been literate. Most parents and children, and most masters and apprentices, would have continued to pass on information by speech and physical demonstration.

RISE AND FALL OF EMPIRES

The city-state of Akkad arose somewhere to the north of Sumer roughly 4300 years ago (2300 BC). Its ruler, Sargon, conquered all of Sumer, all of Mesopotamia, pushed into the Levant, landed in Crete and got as far north as Anatolia, as far east as Elam and as far south as the tip of the Arabian Peninsula. Multiple cultures were incorporated into the Akkadian Empire, and in some cases the Akkadian language was imposed on subjugated peoples. But even this empire lasted until only around 4150 years ago (2150 BC), when it collapsed.

This would not be the last time such a thing would occur.

This is a phenomenon known as a secular cycle, which drives the rise and fall of empires. Around 4200 years ago (2200 BC), droughts, an exhaustion of the soil from overuse, and the increased levels of salt in the soil from short-sighted irrigation techniques seems to have significantly reduced the carrying capacity. This appears to have kicked off a population crisis, where famines were more frequent, uprisings by various cities and aristocrats became more common, and the Akkadian Empire's control over Mesopotamia weakened as the empire shrank. Eventually the empire was destroyed by Gutian 'barbarian' invasions.

Ultimately, there is a relationship between collective learning, carrying capacity and the socio-political stability of an

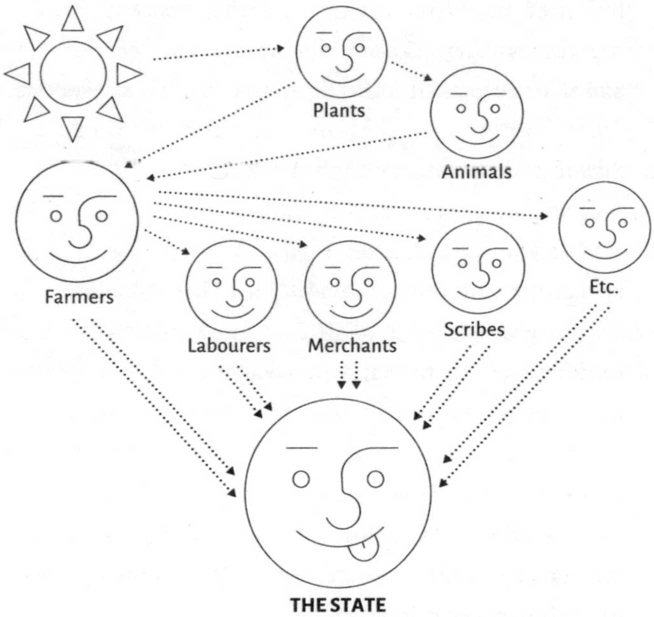
Plants
Animals
Farmers
Labourers
Merchants
Scribes
Etc.
THE STATE

empire. The key to this is that despite the fact that collective learning was gradually increasing the carrying capacity – so that the world population increased from 50 million people 5500 years ago (3500 BC) to 954 million people 200 years ago (1800 AD) – population levels frequently outstripped the carrying capacity.

Agrarian people had so many babies that their innovations in agriculture simply could not keep pace. So, instead, there were cycles of rise and decline every few centuries that had a profound impact on micro-historical events.

The pattern was as follows:

1. **Expansion**: when the population is still low and expanding, things are prosperous for the average person because there is more land, more food and higher wages, the ruling family has fairly good control over the aristocracy, and the empire is generally stable and able to expand its territory.

2. **Strain**: as the population approaches the carrying capacity, the average person pays more for basic essentials and gets paid less for their work (if they get paid at all), rents go up, peasants sell off their land because it no longer supports them, and land and wealth coalesce in the hands of the very wealthy and they multiply in number.

3. **Crisis**: when famine or disease or some other disaster reduces the population, the wealthy lose their peasants, their taxpayers and the source of wealth from rents and payments for agricultural produce.

4. **Depression**; the wealthy begin competing with each other in tremendous civil wars and uprisings, and also competing with the government, until 1) an invading army takes over, 2) the population of the elite is reduced to the point that peace and stability reigns again for another population recovery, or 3) the empire collapses completely and the region becomes depopulated.

Collective learning gradually raises the carrying capacity, but this does not keep pace with population growth, so a kingdom or empire is thrown into these cycles of rise and fall every few centuries. This is how some of the big trends we've been observing have an influence on small-scale historical events.

It is also how humans differ from other species in nature. Usually when a species hits the carrying capacity of its ecosystem, the population crashes and then quickly recovers when the small number of survivors enjoy more food. But in the human case, there is an extra layer of complexity where large-scale violence and civil war can hold a population low for decades after the population crash.

Across the ancient world, we see this pattern playing out – in Mesopotamia, in the Egyptian Old, Middle and New Kingdoms, and in China with the Xia, Shang and Zhou dynasties. The collapse of each is led by a period of population strain, disease and civil infighting, and is often finished off by foreign invasion and occasionally by a brief 'dark age' where history grows quiet.

Almost every civil war, state collapse and era of prosperity and imperial expansion has some connection to this pattern from 3000 BC to 1800 AD (and longer, where agrarian states did not quickly industrialise).

PHASE	POPULATION	REAL WAGE	ELITE NUMBERS	VIOLENCE LEVEL
Expansion	Growing	High, with good standards of living for common people	Low/Moderate	Low, with large amount of state stability
Strain	Slowing	Shrinking, with declining standards of living for common people	Increasing	Rising, with mostly popular rebellions without much elite support
Crisis	Dropping	Increasing, with declining standards of living for common people who survive	Top-heavy societal hierarchy, lower orders begin to be impoverished	Substantially increasing with elite factionalism, competition, discontent
Depression	Kept low	Increasing, with standards of living eroded by violence and possible oppression	Gradually declining as socio-political strife continues	High, with elites competing with each other and weakening government for what resources there are left
Recovery (aka another expansion)	Growing	High, with good standards of living for common people	Low/Moderate	Low, with large amount of state stability

COMPLEXITY IN AGRARIAN STATES

In terms of structural intricacy (the number and diversity of building blocks, networks and connections in a system), agrarian states represent a huge leap forward in complexity. Instead of groups of a few dozen foragers, or early farming communities of a few hundred farmers, there were now cities of tens of thousands, with people doing a dramatic range of different jobs other than farming (a greater diversity of building blocks). These people were increasingly connected in states and empires composed of millions. Between these states, trade routes were growing stronger and more numerous.

In terms of energy flows, we can also see complexity increasing. As in Early Agrarian societies, most of the energy comes from the Sun. Plants absorb this energy via photosynthesis and these plants are then eaten by humans and animals (which humans also eat or from which humans harness their energy for labour). The food and wealth generated from farming goes to support non-farmers (artisans, scribes, soldiers, merchants, cooks, architects, kings and so on) in the rest of the agrarian civilisation.

At the highest level, at the government of the state, much of the energy flows from all this farming and economic activity is taken in the form of rents, tributes and taxation. Currency itself is representative of the energy flows, because currency represents value, and can be used for goods and services. So in order to conduct the complex business of the state, governments harness a greater density of energy flow (average 100,000 erg/g/s) than has been seen before in foraging or early farming societies, or anything else in the Universe.

One can compare an agrarian state to an organism. The latter seeks out food (energy) to sustain or increase its complexity. In the same way, an agrarian state seeks land and wealth. Both organisms and states also compete for these resources. And if either runs out of energy flows, they die. The fossils of animals, the skeletons of humans and the ruins of ancient civilisations all have that in common. They were something great but are no more. They are symbolic of the endgame of the Second Law of Thermodynamics.

EVOLUTION OF AGRARIAN STATES

Between 5500 years ago (3500 BC) and 2000 years ago (circa 1 AD), the world population grew from 50 million to 250 million people. Some 90 per cent of them existed in Afro-Eurasia, 8 per cent of them existed in the Americas, and 2 per cent existed in Australasia and the Pacific. 5500 years ago (3500 BC) the city-states of Mesopotamia and the kingdom of Egypt only controlled 0.2 per cent of the Earth's terrestrial surface. By the time East Asia, West Africa and the Americas produced their first Agrarian States (the last of these rising in the Americas about 3000 years ago), that percentage grew to 6 per cent. By 1000 AD, the land area controlled by Agrarian States had increased to 13 per cent. The vast majority of the Earth's land was populated by stateless farmers or foragers, or was uninhabited by humans.

The world at this time can be divided into four zones: Afro-Eurasia, the Americas, Australasia and the Pacific. These divisions are based on collective learning. No collective learning was passed between these world zones before the so-called Age of Explorations and their unification in one single web of collective learning. But there was information exchange between states and peoples *within* the Americas, Australasia or the Pacific. These same goes for the continents of Africa, Europe and Asia (though long-distance exchange could take several generations), which is why they are grouped into the Afro-Eurasian world zone.

Afro-Eurasia had the major advantage in collective learning. It had the largest populations – for instance, the large and sprawling Achaemenid Empire in 480 BC contained an estimated 50 million people, or 40 per cent of the world population at the time. Agriculture had first arisen in Afro-Eurasia, as had agrarian states, so it makes sense that the largest clusters of populations existed in East Asia, India, the Mediterranean and West Africa, rather than elsewhere in the world. It is here we see the rise and fall of the various Chinese dynastic empires, the Persian, Greek and Roman empires, the origin and disappearance of the Indus River Valley civilisation and the gold-rich states of Mali.

The vast masses of millions of people in Afro-Eurasia prompted the evolution of diseases. Agrarian states were not any more sanitary than Early Agrarian societies (where people were sedentary, lived at close quarters with livestock and drank contaminated water) and the larger populations allowed diseases more opportunity to evolve in deadlier and deadlier forms. Smallpox, bubonic plague and a cocktail of other diseases swept across Afro-Eurasia multiple times in the Agrarian Era. This world zone acting as a petri dish for human contagion was to have a grim impact on the Americas, Australasia and the Pacific with the unification of the world zones.

The Americas first adopted agriculture around 5000 years ago (circa 3000 BC) and the first states emerged in Mesoamerica around 3000 years ago (circa 1000 BC). There was slight lag behind Afro-Eurasia in agriculture, which is why the American world zone had only 8 per cent of the world population. Nevertheless, the *cordon sanitaire* of the Atlantic allowed the human experiment to run independently and much the same results occurred. In 500 AD, the city of Teotihuacan hosted a population of nearly 200,000 people, which is massive by any pre-modern standard. The agrarian states of the Olmecs, Mayans, Aztecs and (further south) Incas all carry the trappings of advanced civilisations.

THE WORLD BEYOND AGRARIAN STATES

Large tracts of Northern Europe, the Sahara and Arabian deserts, and the plains of Central Asia remained stateless for many centuries – populated either by Early Agrarian societies or nomadic foragers. These hinterlands posed a major threat for agrarian states, should the states weaken during a crisis or depression phase of a secular cycle. It is not a coincidence that so many Chinese dynasties, for instance, were founded by 'barbarian' invaders, or that the Germanic invasions supplanted the Roman Empire in Europe.

Central and South Africa, from the Sahara to the Cape, resisted agriculture and states for a much longer period. The delay in agriculture is because Sub-Saharan Africa is one of the best environments for human foraging and one of the least forgiving for sedentary farming. Nevertheless, agriculture had spread into Central Africa by 1500 BC and managed to pierce deep into the Congo, with some cultures adopting agriculture by 500 BC. The practice of agriculture reached southern Africa by 300 AD. A few agrarian states began to appear in these regions just prior to the unification of the world zones circa 1500 AD.

North America saw Early Agrarian societies emerge by 600 AD, most notably the Pueblo societies of the US south-west. The most impressive settlements that arguably could be called an agrarian state were in Chaco Canyon, which held 5000 people and was built between 850 and 1150 AD. Beyond in the Great Plains, California, the eastern seaboard and Canada, semi-sedentary cultures mixed farming and foraging, while some exclusively foraged. But for the arrival of Europeans, agrarian states would probably have arisen in these regions as well.

In the Australasian world zone, the people escaped the trap of sedentism and the unsanitary conditions of agriculture completely. In terms of health, it was certainly far more preferable for people to forage, especially because Aboriginal practices were so productive. Aboriginal people used fire to burn down forests, kill game and expose edible plants, and to clear paths for travel, while the fire-loving eucalyptus forests quickly recovered. The Australian continent was able to support somewhere between an impressive 500,000 and 1 million foragers.

The Pacific world zone was only inhabited by humans in the last 5000 years, and some islands less than 2000 years ago. New Zealand was only settled in 1280 AD due to the lack of sailing winds blowing down from the north. The populations in this world zone consisted of islands of a few hundred people through to larger island chains that contained thousands. For instance, the Hawaiian Islands supported up to 30,000 people, and there was a degree of domestication and irrigation there that arguably could be called agriculture.

THE SILK ROADS

There were many agrarian civilisations spread across Afro-Eurasia who could potentially share collective learning. But often these states were separated by vast distances, huge deserts or impenetrable forest, and travellers might be captured or killed on their arduous journeys. So for 3000 years, transmission of collective learning was slow. The first trade route that crossed the entire Afro-Eurasian world zone did not exist until 50 BC. The Silk Roads allowed slow transmission of goods and information from China to India to Persia to the Mediterranean and, via the Saharan trade routes, to West Africa.

Despite the name, the Silk Roads did not just transmit silk, spices or other commercial goods. They also transmitted religions, inventions and mathematical ideas. For instance, Hindu numerals were invented in the 400s AD, were picked up by the Arabs during the Islamic invasions (hence the misnomer 'Arabic numerals') and made their way into Europe in the Middle Ages, replacing the clunkier system of Roman numerals.

Due to its large populations, many luxury goods and spices, China was the centre of gravity for trade along the Silk Roads. Chinese goods were slowly transported piecemeal across Central Asia, often by nomads, and often taking more than one generation, but ultimately flooded into Middle Eastern and Mediterranean markets. In return, the West gave the East grapes, manufactured products and horses. But the balance of trade lay with the larger populations of Asia.

The land-based route of the Silk Roads was an agonising journey from East Mediterranean ports across the grit of Mesopotamia and Persia, and over numerous mountain ranges and deserts into India and China. These Central Asian routes not only meant a harsh commute but engagement with numerous nomadic and imperial forces, who might kill or rob you. Otherwise, the sea route ran down the Red Sea to Aksum (which made a fortune, allowing its tiny population to become a mercantile superpower in the first millennium BC), on to one of the many ports of India and then onwards to Indochina and South China. It was by these sea routes that Islam spread from India to Malaya and Indonesia.

Afro-Eurasia had a conduit through which collective learning could flow through hundreds of different agrarian states on a super-continent which in 1000 AD was populated by roughly 300 million people. The overwhelming majority of trade was not carried out by one merchant from one end of the Silk Roads to another, and the transmission of trade goods and information could take years or even generations to travel across Afro-Eurasia. Nevertheless, the Silk Roads were kicking off a slow, creeping revolution. Contemporaries at the time would not have noticed this, but it would soon provoke massive changes in human history.

THE EVOLUTION OF PRINTING

In the pre-modern world, the greatest limitation to written knowledge was its circulation. The vast amount of collective learning still occurred orally in the pre-modern period, with all the slowness and flaws that method produces. Literacy still remained in the hands of scribes, bureaucrats, philosophers and the elite. Written works remained scarce and expensive. Printing would change all that.

Original Chinese printing was done with woodblocks, emerging at the end of the Han period in 220 AD. Each page had to be carved into the blocks, slowing efficiency. Woodblocks were extremely bulky and hard to store and transport, every new copy or variation being started from scratch. In 1045 AD, Bi Sheng invented movable type, where words were placed onto clay tablets which could be rearranged to create a new sequence of words and then be imprinted on the page. Numerous Chinese philosophical, scientific and agricultural works were regularly produced. Some books had a circulation of thousands.

The Koreans in the 1200s AD invented a metal moveable type. The advantages of metal moveable type are that it is durable, smaller and easier to rearrange, and ultimately produces books at a faster rate. The Koreans did not use a press of any kind. They laid thin paper over inked type and took the impression by rubbing it with a wooden spatula. This was painfully slow. Nevertheless, with woodblocks or metal moveable type applied by spatulas, the impact of early printing in China and Korea was to reproduce more copies of written knowledge at a faster rate than could be done by hand, increasing the number of books in circulation and the array of knowledge that could reach any one person who was literate. It must be noted that in East Asia, slower woodblock printing remained the dominant form until the nineteenth century, setting a limit on the collective learning that could be derived from the wider circulation of works.

In Europe, Gutenberg's printing press was invented around 1450 AD and combined metal moveable type (imported via the Silk Roads from the East) with a wine press (one of alcohol's many virtuous usages) for quick composition of new pages and then relatively fast imprinting upon paper. This revolutionised printing. In the 1460s, three men could make 200 copies of a book in 100 days with a Gutenberg-style printing press. The same number of copies would take three medieval scribes thirty years. During the sixth century AD, Benedictine monasteries made a rule of housing around fifty books. The largest library in the mid-fifteenth century West was that of the Vatican. It contained around 2000 books. A private scholar of middling rank could easily acquire that many in the seventeenth or eighteenth century.

There were an estimated 8 million copies of books produced in the fifty years between 1450 and 1500 AD. This quite likely exceeds the entire number of books that were hand-copied in Europe since the year 500 AD. Between 1500 and 1600 AD, between 140 and 200 million books were printed. This proffered the Europeans a tremendous advantage in collective learning that aided the spread of the Renaissance and Reformation and would trigger the Scientific Revolution.

Greater profusion of information, greater connectivity and a gradually rising rate of literacy meant that another explosion in complexity was just around the corner.

10

THE UNIFICATION OF THE WORLD

Wherein Afro-Eurasians stray into other world zones • China nearly launches the Industrial Revolution • An Afro-Eurasian disease in a long line of diseases kills millions • The Turks inadvertently cause the next rise of complexity • Slavery continues • And the costs of complexity are made palpable by and for humanity.

BY 1200 AD THE WORLD POPULATION was approximately 400 million, but not without the dips and declines of secular cycles. For instance, in 1 AD the world population was roughly 250 million. But after the decline and collapse of the Roman Empire, Han China and numerous other agrarian states, the world population had shrunk to 200 million by 600 AD. By 1200 AD, the world had recovered and massively surpassed its ancient maximum.

The land-based and seaborne Silk Roads had unified Afro-Eurasia into a fairly robust network of collective learning with shared ideas and innovations (and, horrifically, diseases). The other three world zones – the Americas, Australasia and the Pacific – had not yet entered this network to spur on collective learning to be even faster. The unification of the world zones into a single web of collective learning would harness the innovative power of all humans on Earth and set humanity on a collision course for modernity, with all the higher complexity it has brought.

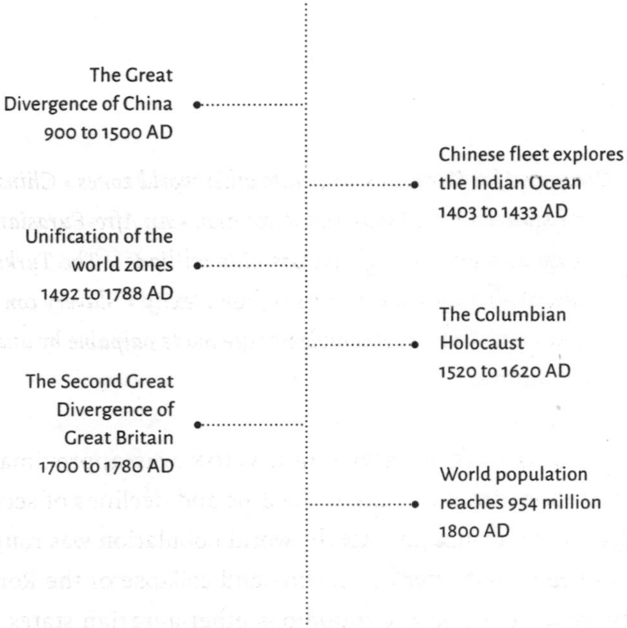
The Great Divergence of China 900 to 1500 AD
Chinese fleet explores the Indian Ocean 1403 to 1433 AD
Unification of the world zones 1492 to 1788 AD
The Columbian Holocaust 1520 to 1620 AD
The Second Great Divergence of Great Britain 1700 to 1780 AD
World population reaches 954 million 1800 AD

THE MEDIEVAL ORIGINS OF GLOBALISATION

An agrarian state gained the overwhelming majority of its wealth (that is, energy flows) from agriculture. Landlords would have a share of the crops produced, or would collect rents, and central governments would collect taxes and tributes. But thanks to the Silk Roads, that was changing. Merchants were making more money and were becoming increasingly influential. The Italian merchant states of Venice, Genoa and Florence had become the wealthiest in Europe despite their small size, the spice traders and Tamil kings of Ceylon and South India managed to achieve the same dynamic, and the bustling international spice trade brought increasing wealth and power to the Srivijaya kingdom in Indonesia. All were smaller states that wielded tremendous wealth simply by account of commerce, which in many ways outstripped states of a similar size that relied on tax income from the land.

In the eleventh century, the start of the Crusades brought Europe into closer contact with the Middle East. The Vikings made temporary forays into North America. In 1271 AD, Marco Polo travelled across Central Asia on a perilous journey to China, and when he published his account of these travels in 1300 AD, European society was shocked at the level of riches in East Asia, intensifying the motivation of European merchants to trade there.

And there was a very good reason to be motivated to trade in China. Equal interest did not exist along the Silk Roads. Europe and Africa were eager to access Asian markets in order to obtain what they could not produce themselves – silks, spices, pottery and so on. The successive Muslim caliphates of the Middle East were still making quite a packet of cash playing middlemen for Chinese goods and Indian spices, since they dwelt along the only trade routes to the East (nobody had rounded the Cape of Africa yet). China, however, produced the most desirable goods and in large quantities and thus had a dominant position in the entire network of trade. Just as modern globalisation has been driven by a wealthy and technologically advanced West, the medieval origins of globalisation were driven by a wealthy and technologically advanced China.

THE FIRST GREAT DIVERGENCE

Long before the West raced ahead of the rest of the world economically and technologically in the nineteenth century, China had done the same centuries earlier, to the point that it could have had an industrial revolution. This was entirely down to collective learning, driven by the number of potential innovators, people who could come up with a new innovation in their lifetimes. The larger the population of humans, the more rolls one has at the dice every generation.

Between 500 and 1100 AD, with the spread of wet rice production in southern China, the carrying capacity absolutely exploded. Per hectare, traditional varieties of rice support around six people compared to three people on a hectare of wheat. This intensified in the Song dynasty (960–1276 AD). The Song government introduced a high-yield strain of rice from Vietnam. They appointed 'master farmers' from local communities, and disseminated new farming techniques and knowledge of new tools, fertilisers and irrigation methods. The Song also introduced tax breaks on newly reclaimed land and low-interest loans for farmers to invest in new agricultural equipment and crops. The Chinese government distributed 3000 printed copies of *Essentials of Agriculture and Sericulture* to landowners in order to improve crop yields. Wet rice farming by this method produced two to three crops a year.

During the 900s and 1000s AD under the Song dynasty, the carrying capacity of China increased from 50 to 60 million to 110 to 120 million (nearly half the world's population), with record high population densities, such as 5 million people farming an area of 65 by 80 kilometres. By 1100 AD, China constituted 30 to 40 per cent of the population of the globe, compared to Europe, which constituted 10 to 12 per cent.

Chinese collective learning advanced by leaps and bounds. For instance, the annual minting and use of coin currency was increased greatly under the Song dynasty. They introduced paper currency. Farming techniques improved: use of manure became more frequent, new strains of seed were developed, hydraulic and irrigation techniques improved, and farms shifted to crop specialisation. Coal was used to manufacture iron (the same iron process that fuelled early British industrialisation), and production increased from 19,000 tonnes per year under the Tang (618–907 AD) to 113,000 tonnes under the Song. The Song dynasty was the first to invent and harness the power of gunpowder. Textile production showed the first ever signs of mechanisation.

It is interesting to speculate what modern history would have looked like if the Industrial Revolution had happened in

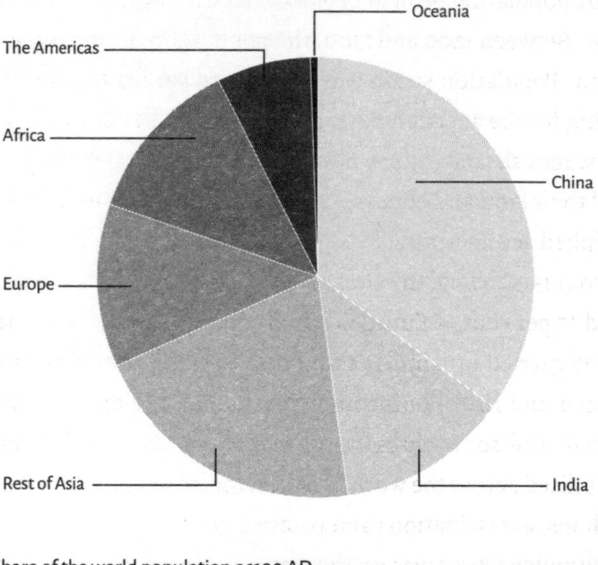
Oceania
The Americas
Africa
China
Europe
Rest of Asia
India

Share of the world population c.1100 AD

China. Certainly the socio-political history of the world would have looked decidedly different, with Chinese ships arriving off the coasts of America and Australia, perhaps with the intent of colonising there (and inadvertently passing on deadly Afro-Eurasian diseases), and the Age of Imperialism happening at the expense, rather than to the profit, of Europe.

THE BLACK DEATH AND SECULAR CYCLES

Thanks to collective learning, innovations in agriculture and the opening up of new lands to farming, the world population ballooned from 300 million in 1100 to 400 million by 1200. The common population was fairly prosperous, with low rents and decent wages, ate meat semi-regularly and were in fairly rude health by pre-modern standards. Agrarian civilisations were fairly stable (internally at least), compared to the periods that came before and after.

But population growth began to outstrip agricultural innovation. Between 1200 and 1300, the population only grew to 432 million. Population strain was starting to set in. The standard of living for the agrarian peasantry declined: they ate less meat, their wages shrank, their rents rose, smallholding farmers had to sell their land and elites set up huge estates as their numbers multiplied several times.

From 1315 to 1317, the Great Famine of Europe killed an estimated 15 per cent of Europe's population. A famine in China in 1333–37 carried off similar numbers. The common population declined and the elite incomes that depended on them were reduced, and some elites even became impoverished. Politics destabilised across the world, with an increased number of elite rebellions, assassinations and palace coups.

But much worse was on the way, courtesy of the Silk Roads.

The Black Death was an extremely deadly disease caused by the bacterium *Yersinia pestis*, carried by fleas, which in turn are carried by rats. Once bitten by an infected flea, a human develops swollen lymph nodes around the groin, which are painful at the touch. As the bacteria works its way through the human bloodstream, plague victims experience fever, weakness, delirium, headache, bloody vomiting and the death of flesh and internal organs, which can turn black and gangrenous. Death usually occurs within a week to ten days, and death was the outcome for roughly 80 per cent of bubonic plague victims. If the plague turned pneumonic, the death rates was between 90 and 95 per cent; very often people with pneumonic plague died within just two to three hours of catching it.

The Silk Roads transported the Black Death east and west. Sporadic outbreaks occurred in China in the 1340s and a more severe pandemic swept across all of China between 1353 and 1354. The resulting population downturn, elite infighting and state collapse led to the overthrow of the Yuan dynasty in favour of the Ming in 1368. Plague and a depression phase reduced the Chinese population from 120 to 140 million in 1200 to 65 million people by 1393.

The Black Death also spread to Persia from Central Asia by 1335, killing an estimated 30 to 50 per cent of the population, including the ruler of the Ilkhanate, which had dominated the region since the fragmentation of the Mongol Empire in 1256–59. This caused the collapse of the khanate into several rival kingdoms. From 1338 to 1344, the Black Death spread across the trade routes of the Golden Horde to the north, killing an estimated 30 to 70 per cent of the population.

In 1344, an already infected army of the Golden Horde besieged the Genovese-held Crimean trade port of Kaffa. The

Golden Horde placed plague corpses onto catapults and flung the bodies over the city walls, in one of the first recorded instances of biological warfare in human history.

The Black Death managed to get on board Genovese trading ships and began to infect the Mediterranean. The disease arrived in the port of Constantinople in 1347 and then spread overland across Anatolia, reaching Damascus in 1348, where an estimated 2000 people died per day. That same year the Black Death reached Egypt, where it killed an estimated 50 per cent of the population of Cairo. Due to the Muslim tradition of making a pilgrimage to Mecca, the Black Death arrived in their holiest of cities in 1349.

Back in 1347, the Genovese traders had also travelled on to Greece, Sicily, Sardinia, Corsica and Marseille. In 1348, traders reached England, Ireland and northern France. In 1349, the Black Death rushed through southern Spain and got as far as Morocco. Infected ships arrived at Bergen in Norway. The plague spread from England to Scotland, from Norway to Sweden, and from France to the Holy Roman Empire in 1350. Poland and Russia were hit in 1351 and 1353. Only the small, cold-dwelling population of Finland was spared.

In 1300, the world population had been 432 million. By 1400, from a mixture of famine and plague and the decades of violent instability that usually follow depopulation, the world population had shrunk to 350 million people.

An intriguing side-effect of population collapse, however, is that life becomes quite prosperous again for the common people. A shortage of labour means higher wages, an abundance of land left over by the dead means lower rents and sustainable peasant farms, while food is cheap because there are fewer people buying it. The peasantry might even possess what we would call a 'disposable income' for a few modest luxuries. The common people of Afro-Eurasia had a high standard of living and a larger 'real wage' than in any period prior to the Industrial Revolution.

Dr. Doctor Schna- -bel von Rom
Kleidung wider den Tod zu Rom. Anno 1656.

Plague doctor

THE AGE OF CHINESE EXPLORATIONS

After the Black Death, the Ottoman Turks shut down much of the overland trade on the Silk Roads. The disruption of the network of exchange prompted explorers at either end of the Afro-Eurasian supercontinent to find new routes by sea.

In 1403, the Ming dynasty began construction of a gigantic fleet of warships and merchant vessels that dwarfed anything else in the world at the time. The Chinese exploration fleet was composed of 317 ships, some of them approximately 120 metres high, with three or four decks, carrying a massive army of 28,000 men to, shall we say, add weight to trade negotiations.

Starting in 1405, numerous expeditions were launched. The Chinese fleet sailed around South-East Asia and to India on several occasions. They sailed down and conducted trade in Indonesia. They made landfall in Arabia and East Africa on several occasions. In total, seven expeditions were made.

The final voyage ended and returned home in 1433. At this point China, already a mighty empire with many natural resources and luxury goods, turned to isolationism. If these voyages had continued, it is not unreasonable to speculate that the Chinese could have eventually sailed around the southern tip of Africa, perhaps even achieving direct trade with Europe. It is also possible the Chinese could have moved further south from Indonesia to Australia. And perhaps the Chinese could even have navigated the Pacific, taking them to the Americas.

THE AGE OF EUROPEAN EXPLORATIONS

In the fifteenth century, European states could not gain as much income from taxing the farming population as they had prior to the Black Death, and so merchants and commercialism began to be looked upon more favourably. Yet the Ottoman Turks had shut down much of the trade via the Silk Roads in their bid to conquer Europe. The Europeans recoiled West.

By the 1420s, the Portuguese and Spanish had landed in the Canaries, Madeira and the Azores, and charted a good distance down the seemingly unending African continent. In the 1440s and

1450s, Portugal began a great deal of trade with the Mali Empire. Portugal gained access to pepper, ivory, gold and the African slave trade. Bartolomeu Dias reached the Cape of South Africa in 1488. And in 1498, Vasco da Gama rounded Africa and reached India, and brought back a cargo of spices. Bypassing the hostile Ottomans, da Gama was able to purchase his cargo for 5 per cent what it would have cost him buying from eastern Mediterranean sources.

The problem with sailing around Africa is that near the equator one hits a region called the 'doldrums'. This is a long stretch of ocean where the winds are frequently too weak to propel sail, and where ships were frequently hit with dangerous storms.

Alternatives were already being sought. In 1492, Ferdinand of Aragon and Isabella of Castile commissioned an expedition led by Genoese explorer Christopher Columbus (little did they know the voyage had already been made by the Vikings nearly 500 years earlier). Columbus left Castile in August, sailed west, and reached the Bahamas in October. He went on to visit Cuba and Hispaniola. Upon the native populations, Columbus established a regime of enforced slave labour, sexual slavery and mutilation for disobedience; all the while the island populations were gradually wiped out by European diseases. To his dying day, Columbus was certain that he had landed in Asia.

In 1519, the Spanish monarchy commissioned Portuguese explorer Ferdinand Magellan to take five ships, sail south of the Americas and enter the Pacific, whereupon Magellan crossed the vast ocean and reached the Philippines, where he was killed in 1521. Only one ship managed to return to Spain, in 1522, under command of Juan Sebastian del Cano, being the first crew to circumnavigate the globe.

The sixteenth century set off an explosion of European and colonial merchant ventures to Asia and the Americas, with states, private investors and individuals seeking vast fortunes. Habsburg Spain dominated these trade networks and the acquisition of colonies in some of the most mineral-rich parts of South and Central Americas. Other prominent nations that joined the colonial effort were England, France and the Netherlands, with even a few attempts to set up colonies by Scotland. The nations of Central and Eastern Europe largely missed out on the Age of Explorations due to local wars and their geographic locations.

From 1519 to 1521, Hernán Cortés led a few hundred Spanish conquistadors, armed with gunpowder weapons and a cocktail of diseases against the Aztecs. As a result of the high death rates of Aztecs at the hands of European disease and Cortes allying with various local enemies of the Aztecs, all of Mexico fell into Spanish hands within a few short years. In 1532, Francisco Pizarro led a similar expedition against the Incan Empire, and was also aided by gunpowder weapons and the horrible ravages of European disease. However, the Incan Empire stretched across vast and difficult terrain and the Spanish did not fully conquer the empire until 1572, as the result of a long and gruelling war of annihilation.

THE SLAVE TRADE

In the Caribbean and South America, Europeans had stumbled across a climate that was ideal for sugar plantations. The question was finding enough labour to engage in such gruelling work. Lower-class Europeans weren't the answer. Only indentured servants shipped over to the Americas could possibly be forced into it. And they soon moved on after their contract was up, and there certainly weren't the numbers of people willing to do it. The Spanish and Portuguese initially tried forcing the native inhabitants of the Americas to do it. But they knew the countryside and frequently escaped back to their people. Those who stayed often died of Afro-Eurasian diseases. So the Portuguese utilised the slave trade with African rulers they had entered into half a century earlier.

Slavery had existed since the inception of agrarian states 5500 years ago. There had been slavery in Europe, in Africa and in Asia. The Aztecs and the Incans had held slaves. As had the Chinese, Koreans and Indians. Of the roughly 55 billion people who lived in the Agrarian Era, an estimated 3 to 10 billion of them may have been slaves.

Europeans were no strangers to slavery. The Romans had owned huge plantations across the Mediterranean world operated by millions of slaves. The Middle Ages blurred the lines between slavery and serfdom, which wasn't much better (though it certainly was an improvement). In fact, serfdom was an early medieval perversion of older systems of slavery, even in its name: 'serf' being derived from the Latin *servus* (slave). To the east, Russia did not abolish serfdom till 1861.

The kingdoms of West Africa had been engaging in the Muslim slave trade for several centuries by the fifteenth century, with the forced transport of people across the Sahara. The demand for African slaves had increased since the number of Europeans captured and enslaved by Muslims had slowly declined since the eleventh century. The Africans themselves primarily derived their slaves from peoples they conquered in war (but debt or birth into a slave kin-group were also causes for slavery), and Africans either kept them or sold them off across the Silk Roads. When the Portuguese opened up trading relations with African rulers in the 1440s, the slave trade was extended to them as well.

The horrors of transportation killed 10 to 20 per cent of Africans moved across the Atlantic. The route-march east across the Sahara claimed 25 to 50 per cent of those sold or captured. In total, 11 to 14 million Africans were taken west across the Atlantic in just 400 years, and 10 to 17 million Africans were taken east across the Sahara over 1100 years. An average of 5 to 15 per cent of the population were slaves in the agrarian states of Africa.

Slavery was the rule of agrarian states. Absence of slavery was the exception. From the perspective of the chained, across the globe, it had been a hideous 5000 years.

The Portuguese were responsible for 45 per cent of all slaves taken during the Atlantic slave trade. Their former colony, Brazil, was the destination of 35 per cent of all slaves in the Atlantic trade and also one of the last of these countries to abolish slavery, in 1888. The Spanish account, for roughly 15 per cent of all African slaves, transporting most of them to South America and their Caribbean island holdings. They also made more determined use of Native American slavery, particularly in their mining operations. The French transported 10 per cent of African slaves to their holdings in the Caribbean, mostly plantations. The Dutch did the same with 5 per cent of the total number.

In the seventeenth and eighteenth century, forced plantation slave labour expanded from sugar to include the production of tobacco (another highly addictive product) and cotton for textiles. This made slave labour desirable for farms in the southern half of the Thirteen Colonies. As such, the British imported 15 per cent of the total number of African slaves to their Caribbean plantations and approximately 10 per cent were imported to lives of bondage in the future United States, for a total of 25 per cent of the slave trade.

In the 1500s, an estimated 400,000 to 500,000 Africans were enslaved by Europeans (1 per cent of Africa's population). In the 1600s, this number increased to 1 to 1.5 million (2.5 per cent). In the 1700s, 5 to 8 million people (10 per cent) were bought, bound, packed into appalling conditions in ships and sent to the Americas.

It was the horrific scale of the 1700s that finally sparked the abolitionist movement in Great Britain. A thirty-year public and parliamentary campaign led to the British abolition of the slave trade in 1807, making further purchase and transportation illegal. And the British Navy became actively engaged stopping the transportation of African slaves by other nations. Nevertheless, the remaining slaver Atlantic states succeeded in transporting an additional 3 to 4 million slaves from Africa in the 1800s (4–5 per cent of Africa's population). The British Empire abolished slavery itself in 1833, and slowly, either by brutal civil war, gruesome revolution or peaceful legislation, the rest of the Atlantic nations followed suit in the subsequent decades.

The African slave trade

In Africa itself, slavery continued, particularly in North Africa, where slavery was given a religious and ethnic justification. In the late nineteenth century, European imperialism and intervention attempted to expunge slavery from Africa itself, but very slowly and often ineffectively. Or, depending on the colonial power, even half-heartedly. Or prolonged it.

Even today, in the postcolonial era, slavery in Africa is still a problem. In modern Nigeria, there are 700,000; in Ethiopia, 650,000; in the Congo, 500,000; and in total roughly 5 to 10 million people leading lives effectively as slaves in Africa. Beyond Africa, there are 12 to 14 million de facto slaves in India, 2 million in Pakistan and 3 million in China. There are currently 47 million slaves worldwide, which is roughly the population of Spain.

ECOLOGICAL IMPERIALISM

Europeans brought to the Americas and Australasia all their domesticated farm animals, a necessary ingredient for settler colonies. Sheep and cattle were bred so prodigiously that they quickly became some of the most common mammals to be found in either world zone. In the Americas, by 1600 there were 20 million sheep and cattle.

When humans arrived in the Americas 12,000 years ago, they hunted American horse species to extinction. Once Europeans arrived, they reintroduced the horse. Some horses were obtained by Native Americans. As a result, the lifeways of Native Americans in the Great Plains were radically changed. Many cultures switched from agrarian cultures that had existed in the Great Plains for thousands of years to become nomadic foragers once again. Prior to the horse, Native Americans would camouflage themselves with pelts and crawl along the ground to get close enough to herds of buffalo. When they got near, they'd spear a buffalo before the herd started to stampede away. With the arrival of the horse, Native Americans could now keep pace with the buffalo and spear them as they chased them, or else herd them off the edge of cliffs. By the nineteenth century, the Great Plains peoples had kept horses as a keystone of their culture for 300 years. Long enough that some Native American accounts depict the horse as always being in the Americas and part of their way of life.

New World crops in turn impacted Afro-Eurasia. In terms of caloric intake, maize bequeaths more than wheat and only falls short of rice per square kilometre. An example would be potatoes, which is not only quite good on the caloric front but actually enriches the soil with nutrients as they grow. Maize and potatoes also come with the advantage that they are easier to prepare and cook than something like wheat or rice. The Americas also gave the world tomatoes, yams and squashes that were similarly high-yield per square kilometre. The European carrying capacity, where American crops were adopted, increased by 20 to 30 per cent. In China, the population was starting to experience mass famines in the 1630s, but adoption of American crops prevented another mass famine from happening until the nineteenth century, during which time the Chinese population increased from 150 to 330 million people.

Disease utterly destroyed the inhabitants of the Americas and Australasia. For thousands of years, at this point, Afro-Eurasia had been host to 90 per cent of the human population, most of whom were living in densely packed agrarian states where basic hygiene and germ theory were conspicuously absent. Inhabitants of Afro-Eurasia had developed genetic resistances to diseases over hundreds of generations, but the populations of the Americas and Australia had no such biological resistance. European colonists brought over smallpox, typhoid, cholera, measles, tuberculosis, whooping cough and numerous influenzas. While these were still deadly to Europeans, the impact on indigenous populations with no resistance was much worse.

In the Americas, Afro-Eurasian diseases are estimated to have wiped out 90 per cent of the population between 1500 and 1620. Some 50 million people were killed in a world populated by 500 to 580 million. Just by the simple act of Europeans showing up. And that was done in the space of a century. In 1620, only 5 million Native Americans remained in North and South America. This was an unsurpassed obliteration of human civilisation the equal of which cannot be found in human history. Afro-Eurasian diseases would continue to wreak havoc on Native American populations throughout the nineteenth and into the twentieth century.

In Australia, the cocktail of Afro-Eurasian diseases reduced the Aboriginal Australian population by at least 73.75 per cent between 1788 and 1900. Most scholars today agree on a pre-contact population of around 800,000. In 1850, the population had already shrunk to only 200,000 people. In 1900, the Aboriginal population was 90,000. Using Butlin's calculations, that European expansion of farmland caused an upper limit of 100,000 Aboriginal Australians to starve to death, and using Reynolds' estimate for the unrecorded number of Aboriginals who died in frontier violence against recorded European deaths, that makes 73.75 per cent of the population reduction being from disease, 12.5 per cent from starvation and 2.5 per cent from frontier violence in a little over a century.

Imagine if, in the space of a century, 8 billion people on Earth started dying until there was only 800 million left. Cut down the population number of your own nation to 10 per cent if that strikes closer to home. That was the cost of the unification of the world zones. While American crops were raising the carrying capacities of Europe and Asia, sending their populations soaring, the populations of the Americas and Australasia were being reduced at nightmarish rates. Even writing this now, I can barely conceive of the devastation and suffering wrought by such biological terror.

ON THE VERGE OF THE ANTHROPOCENE

Reflect back on the journey of complexity thus far. The firestorms of the Big Bang and exploding stars. The hellish formation of the Earth. The blood-soaked evolution of species. The propensity of primates for murder. The deprivations and disease of the Agrarian Era. And now this. Complexity is *not* synonymous with 'progress'. All of our comforts and conveniences today have come at a tremendous cost that most people can barely contemplate.

Complexity exists in a Universe beset on all sides by the Second Law of Thermodynamics. With every threshold of complexity has come destruction and, where there is consciousness, suffering. Nothing about our current point in history was predestined. It has not been a routine march towards air-conditioning and iPhones. It has been a struggle. A blind struggle. That struggle continues today. But with a little more foresight.

The fascinating part of this history is that with every increase of complexity comes the mounting possibility of final victory over the Second Law and the tremendous natural burdens we have shouldered for 13.8 billion years.

11

THE ANTHROPOCENE

Wherein the British start burning coal for steam power • Mass production kicks off a blizzard of scientific and economic innovation • The rest of the world endeavours to catch up • The world enters a new geological epoch known as the Anthropocene.

THE INDUSTRIAL REVOLUTION IS ANOTHER remarkable threshold of complexity that leads us to the immense transformations of modernity – whether we are discussing the Cambrian-style explosion of new technologies, revolutions of thought and doctrine, or the radical alteration of the lifeways of every human on this planet. To say nothing of how it opened the door to another geological epoch, the Anthropocene, in which humans are impacting the planet more rapidly and drastically than any single species in 3.8 billion years of life. The Anthropocene is a geological period which follows the Holocene (the period that began at the end of the last ice age). The term is derived from *anthropos*, the Greek for human.

We are now living at unprecedented levels of complexity in the history of the known Universe. In terms of structural intricacy, the unified global system of modernity contains an unprecedented number of people (7.9 billion at time of writing), who are all potential innovators in a system of collective learning. And these human minds are united by almost instantaneous communication, transportation and unprecedented levels of literacy. Sustaining this web of knowledge are immensely intricate networks of trade, supply, laws and energy production, and a wider diversification of labour than ever before. In terms of energy flows, the free energy rate density of society has increased from an average of 100,000 erg/g/s in the Agrarian Era to 500,000 erg/g/s in the industrial nineteenth century, to 2 million erg/g/s in developed societies today.

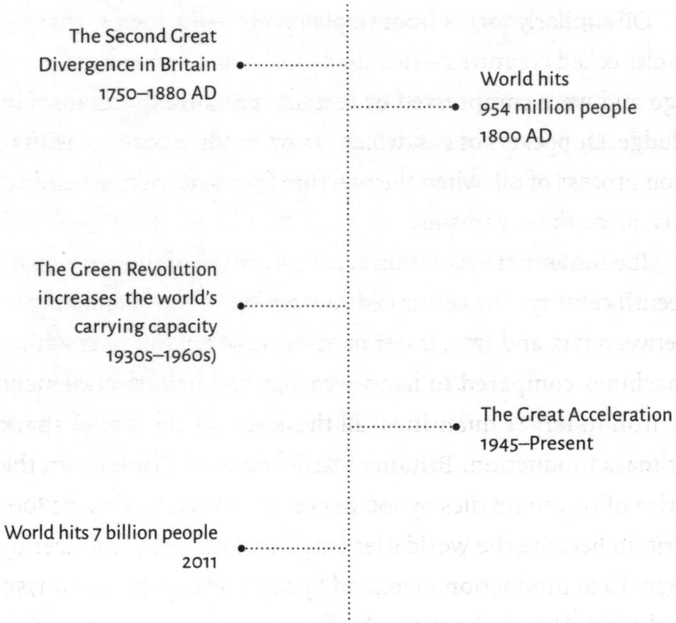
The Second Great
Divergence in Britain
1750–1880 AD
World hits
954 million people
1800 AD
The Green Revolution
increases the world's
carrying capacity
1930s–1960s)
The Great Acceleration
1945–Present
World hits 7 billion people
2011

THE SECOND GREAT DIVERGENCE

The first key ingredient was the harnessing of fossil fuels for industrial production. Fossil fuels include coal, oil and gas. They are so-called because they are actually the remains of living things that perished between 10 and 600 million years ago. Coal is made from the giant trees that fell to the ground starting 350 million years ago and got compressed by plate tectonics to form hard, thick layers of coal in rock beds. When burned, coal released the aggregated energy of billions of plants. Thus, fossil fuels far outstripped the energy output of human labour, animal labour or the burning of wood, when harnessed by industrial machinery. This powered the Industrial Revolution of the eighteenth and nineteenth centuries, and still powers much of the human energy grid today.

Oil similarly forms from single-celled creatures – and some multi-celled creatures – that died hundreds of millions of years ago and were compressed by tectonic pressure into a form of sludge. Or pockets of gas, which is a by-product of the fossilisation process of oil, when the pressure forces out all the residual gas inside the organisms.

The Industrial Revolution began in Great Britain in the eighteenth century. The continued refinement of the steam engine between 1712 and 1775, faster production of textiles in spinning machines compared to hand-weaving, the better refinement of iron in larger quantities: all these set off the initial spark of mass production. Britain's textile industry drove down the price of cotton textiles by 100 per cent between 1750 and 1800. Britain became the world's leading iron and steel producer by 1820. Coal production increased by 600 per cent between 1750 and 1870. Already by 1800, the rate of British manufacturing was three times higher than that in any agrarian state on Earth.

Despite its tiny population, this made Britain the richest nation on the planet.

Farming was no longer as dominant an area of British society. By 1750, roughly 50 per cent of the British economy was built on commercial ventures. Between 1750 and 1850, the number of farmers dropped from 60 per cent of the population to 30 per cent of the British population. A great diversification of labour followed in the nineteenth century, with more specialists such as engineers, lawyers, scientists and entrepreneurs contributing to collective learning than ever before. This set off a further explosion of innovation. A similar phenomenon occurred in every nation that has industrialised.

Britain's lead continued to grow until roughly 1880, when it produced 23 per cent of the world's goods, despite it being tiny in terms of population (approximately 2 to 2.5 per cent of the world's total population in 1880). By comparison, in 1880, China represented 30 per cent of the world's population but only produced 12 per cent of the world's manufactured goods, whereas in 1800 it had produced about 33 per cent of the world's manufactured goods, roughly proportionate to its population size as an agrarian economy.

THE WORLD TRIES TO CATCH UP

The British had a head start on industrialisation over the rest of the world by at least a few decades (in some cases over a century). Britain was able to hold out against Napoleon, his allies and the United States in the Napoleonic Wars and War of 1812, defeated the once mighty China in the First Opium War of 1839–42 and gradually forged the largest empire in human history.

With the advantages of industrialisation becoming more and more apparent, other countries endeavoured to replicate it.

Belgium was already industrialising in the 1820s and 1830s. France began industrialising in the 1840s but with mixed success – only achieving 8 per cent of the world's total manufacturing compared to Britain's 23 per cent in 1880. They caught up later. Kind of. Prussia began industrialising in the 1850s, with the other German states lagging behind, but after unification in 1871, they pursued industrialisation with gusto. Germany would surpass Britain's industrial capacity in the 1910s and '20s. Industrial capacity levels and the incidence of the two world wars are no accident.

The United States was the first power to clearly surpass Britain in terms of industrial production. After the end of the Civil War in 1865, the United States invested in a period of settlement in the west and heavy industrialisation in the north, and let vast numbers of immigrants into the country. By 1880 the population had grown to 50 million, outstripping Britain in numbers, and the United States produced 15 per cent of the world's manufacturing goods. By 1900, the US population had grown to 76 million people, and it produced 25 to 30 per cent of the world's manufacturing. Britain was eclipsed and the United States' lead would only increase.

The modern ingredients for superpower status were clear: have the largest population size possible, provided it was fully industrialised and developed. It is why the nations numbering 1.5 billion people dominate the other 6.5 billion today, and why nations such as China and India continue to industrialise in earnest. Imagine if 1.4 billion Chinese were as industrialised as 330 million Americans today.

Beyond the West were two other early industrialisers eager to even the score and maintain a prominent place on the global stage. Russia tried to industrialise in the nineteenth century, but in 1900 only 5 per cent of the population were industrial workers and the Russian share of total global production was only 8.9 per cent, despite it having a population of 136 million people. It would require World War I, the rise of the Soviet Union and the bloody excesses of Stalin to force a greater degree of industrialisation – and even then the total Russian world share of manufactured goods increased only marginally.

Japan was somewhat more successful. The Japanese embarked on a period of rapid modernisation and industrialisation after the Meiji Restoration of 1868. The central government invited in Western experts, crafted a fairly Western-looking constitution and heavily subsidised all attempts at factory production. By doing so, Japan transformed from a feudal society to a modern one within half a century. Japan had a fairly large population from which they could build a large industrial economy. However, in 1900, Japan made only 2.5 per cent of the world's industrial output, and this share did not grow significantly until after World War II. After that, the Japanese 'economic miracle' allowed its large industrialised population to become incredibly wealthy and it still enjoys the position of third place in the world, after the United States and China, today.

The unification of the world zones, the power of fossil fuels and the imbalance of trade and scientific advancement allowed larger empires than ever before to be forged, and vast swathes of land and the majority of the world's population to be controlled by relatively small armed forces of Europeans, Americans and Japanese. By 1914, roughly 85 per cent of the world's surface had fallen under foreign imperial control.

Two world wars and scores of revolutions have not really altered this imbalance. The United States and Soviet Union wielded immense imperial power (directly and indirectly) in the Cold War. The US has dominated the world stage since 1989. China is currently expanding its influence across Asia, Africa, Europe, Australasia and the Americas with rapid success. Even the French retain an oft-overlooked but significant amount of imperial influence in West Africa. The bulk of the world's nations remain dominated by a few. Anyone who thinks the age of empires ended in the mid-twentieth century ought to look again. Empire is simply conducted more by stealth and with slightly better PR.

THE GREAT ACCELERATION

From 1870 to 1914, the average annual rate of growth of the world's exports was 3.4 per cent and the average annual rate of growth of GDP per capita was 1.3 per cent. The disastrous period of the two world wars from 1914 to 1945 saw the average annual rate of growth of exports shrink to 0.9 per cent and the average annual rate of growth of GDP per capita to 0.91 per cent. Thereafter, the nuclear bomb made warfare between Great Powers too costly. As a mildly ironic result of the creation of the most devastating weapons in world history, the period 1945 to the present has been one of the most 'peaceful' (relatively speaking) in world history for at least 5500 years. Possibly far longer, when one considers the skirmishes and raids of Early Agrarian societies, and the 10 per cent murder rate of foraging societies stretching back 315,000 years to the birth of *Homo sapiens*.

As such, the period 1945 to present has been one of unprecedented growth in terms of exports, GDP, population and complexity. This period is known as the 'Great Acceleration'. From 1945 to 2020, the average annual rate of growth of exports has been 6 per cent and the rate of growth of global GDP has been an average of 3 per cent. Let that settle. Most of the 'busy work' of human complexity has happened in the past seventy years. Still within living memory for some.

Today the United States still holds the lead, with a population of approximately 330 million, producing roughly 25 per cent of the world's GDP. China now has 16 per cent of global GDP from a still-industrialising population of approximately 1.4 billion. Of the next-largest economies, Japan is 5.8 per cent and Germany 4.3 per cent. The much larger Russian population, by contrast, constitutes only 1.8 per cent of global GDP. The combined GDP of the United Kingdom, Australia, Canada and New Zealand is 6.8 per cent of global GDP, which may be of interest should they engage in CANZUK-based unification in the wake of Brexit. India has a large population of 1.35 billion but currently produces only 3.3 per cent of the world's GDP as it lags behind China in industrialisation. In regard to both China and India, the growth of their GDP is really just the adjustment back into proportion with their share of the world population, reversing the Second Great Divergence of the nineteenth century. Provided such enormous populations don't provide an obstacle to further economic growth somewhere in the near future.

Globally, the world population has increased from 2.5 billion in 1945 to 7.9 billion today (although by the time most of you read this book it will probably have reached 8 billion). It took 315,000 years for the world population to achieve its first billion people, it took 100 years for the second billion, and further billions were added every few decades. The Green Revolution from the 1930s to the 1960s produced a number of highly effective chemical fertilisers, pesticides and artificially enhanced grain and rice, raising the global carrying capacity. While regions such as India and China experienced horrific famines in the nineteenth and early to mid-twentieth centuries, their populations have been able to explode since then, soaring into the billions.

The world's GDP output was US$2.7 trillion in 1914, $33.7 trillion in 1997, $63 trillion in 2008, and is $87 trillion at the time of writing. In terms of food production, total grain output has increased from 400 million tonnes in 1900 to over 2 billion tonnes today. The amount of irrigated land increased from 63 million hectares in 1900 to 94 million hectares in 1950 and 260 million hectares today.

Within a very short amount of time, the world has more people producing more stuff than at any point in the last 315,000 years of human history, working in a global system more complex than anything in the past 13.8 billion years. We now live in a network of 7.9 billion potential innovators, within an instantaneous communications network of email and the internet. This bodes well for the acceleration of collective

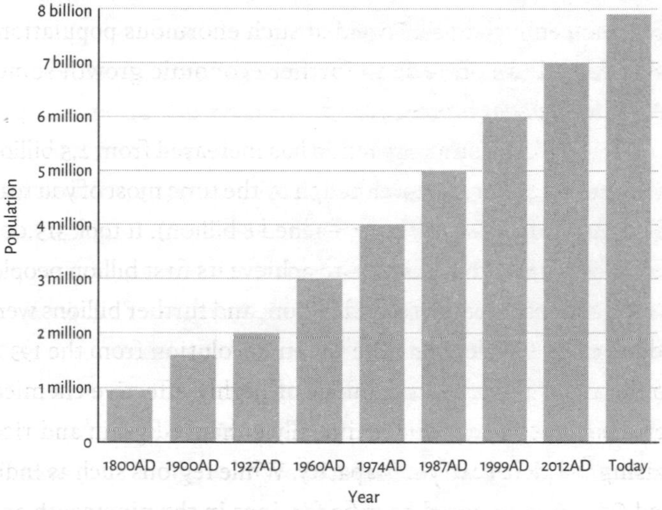
8 billion
7 billion
6 million
5 million
4 million
3 million
2 million
1 million
0
Population
1800AD
1900AD
1927AD
1960AD
1974AD
1987AD
1999AD
2012AD
Today
Year

Population explosion in the Anthropocene

learning into the future, especially as more educational and career opportunities are created for the inhabitants of developing economies.

THE ANTHROPOCENE

By several metrics humans are now the dominant environmental and geological force on the face of the Earth. Not since the Great Oxygenation Event 3 to 2.5 billion years ago have biological organisms had such a profound impact on the Earth's evolution.

There is a debate about when the Anthropocene truly begins. Some date it to the start of agriculture 12,000 years ago, when the immense deforestation from clearing land for farming fields may already have increased carbon emissions, and when humans began terraforming landscapes and domesticating and breeding millions upon millions of new animal species. Most proponents of the Anthropocene concept do not consider these changes significant enough to constitute an entirely new geological era. Others date the start of the Anthropocene to the beginning of the Industrial Revolution circa 1750 or 1800, due to the increase of carbon emissions and the role of technology transforming the environment more than ever before. Others still place the start of the Anthropocene with the Great Acceleration, since most human growth has happened since 1945, and because of the commencement of nuclear weapons testing, which has disrupted the atomic clocks of decaying isotopes around the world.

In terms of pure annual rates of extinction, humans are responsible for an extinction rate as fast as any of the five mass extinctions that have occurred in the past 550 million years, causing some to say that humans are driving a sixth mass extinction in the Anthropocene. Beyond that, human use of fresh water has increased by ten times since 1900, which may threaten to completely dry up the Earth's aquifers, upon which both human and other life depends. We have put 70 per cent of the world's coral reefs at risk. In the past seventy years, we have increased the carbon dioxide content of the atmosphere to over 400 parts per million, higher than anything in the past 3 million years. A lot of this seems to imply a tremendous influence on the Earth system, and none of it seems to bode particularly well.

On the issue of climate change, we have increased the average global temperature by about 1° Celsius since the start of the Industrial Revolution, and we are approaching the same average temperature as the Medieval Warm Period a thousand years ago. If we cross the threshold of more than a 4° increase in the average global temperature, we run the risk of melting the frozen methane stores in the oceans and in Siberia, kicking off a runaway greenhouse effect that could take us to a 5° or 6° increase. In the long term, these increases could reduce arable cropland, starving the population, obliterating still more biodiversity and flooding many highly populous regions with rising seas.

Another concern in the Anthropocene is the sheer growth of the human population. Happily, industrialisation seems to slow down population growth in developed and developing economies alike. Nevertheless, the world population is set to reach 9 billion people by 2050, and somewhere between 10 and 13 billion people by 2100. With most of that population growth happening in regions of the world that are poorest and least equipped to deal with overpopulation – primarily Sub-Saharan Africa. This raises a lot of problems. Either we slow population growth by rapid industrialisation or we don't industrialise (good luck convincing Africa, India and China) – and we run the risk of a Malthusian disaster occurring in regions already nearest to the margins. Already 65 per cent of the world's current global emissions are produced by the developing world. The only long-term solution seems to be technologies such as hydrogen fusion, which would flood the world with cheap sources of energy that had comparatively little environmental impact, so that the world's poor could industrialise and raise their living standards to their heart's content, without risking a global meltdown.

From a bird's eye perspective, it is supremely unsurprising that the first burst of growth following a new threshold of complexity should be followed by a period of strain. We saw it shortly after the adoption of agriculture. We are too early in the Anthropocene to have experienced severe strain. In every stage of evolutionary history, species have exhausted their environments and had to compete for resources and energy flows by adapting their traits. And, ultimately, complexity guzzles all energy flows in the Universe until energy is used up and complexity itself dies.

The question for humanity in the Anthropocene is whether we can innovate in time to avoid another collision with the carrying capacity and another period of horrific decline and mounting death. Whether in this Golden Age we shall ascend to even further heights or descend into an Iron Age of war, or a new Dark Age of obliteration.

That brings us to discussion of the future in the final chapter, in terms of the next few centuries, the next few million years and the next trillion, trillion, trillion years in the life of the Universe.

PART FOUR

THE UNKNOWN PHASE

The present to 10^{40} years from now

12

THE NEAR AND DEEP FUTURE

Wherein humanity's destiny in the Anthropocene is one of four broad possibilities • The natural future of the Universe sees complexity fade • The potential of complexity in the Deep Future may lead to the rise of super-civilisations • The end of the Universe will be either the Big Freeze, Big Rip, Big Crunch or Big Save.

THE UNIVERSE BEGAN AS A WHITE-HOT SPECK of energy within which all the ingredients for everything we see around us – whether by our own eyes or by powerful microscopes and telescopes – already existed. According to the First Law of Thermodynamics, which states that nothing (on the Newtonian scale at least) is created or destroyed but only changes form, we *are* the Universe. Just a highly complex, conscious and self-aware part of it. One totality. Looking at itself. That fact alone is worth celebrating. We have been given the gift, however flawed our vision, of looking into the great beyond. Not many clumps of atoms can claim that honour.

When the physical laws of the Observable Universe became coherent 10^{-35} seconds after the Big Bang, so too emerged tiny dots of unequally distributed energy. The Second Law of

Thermodynamics began to force these dots to even out by sending energy from where there is *more* to where there is *less* in order to achieve a cosmos with equally distributed energy. Energy flows created stars, diverse chemicals, organisms and societies. All complexity in the Universe was created, sustained and increased in complexity by energy flows. From sunlight to photosynthesising plants. From table to mouth. From gas pump to jet engine. In a Universe that is 99.9999999999999 per cent dead, tiny dots in the cosmos have been getting progressively more complex. Regardless of where things go from here, we are lucky to be along for this part of the ride, where complexity is higher than at any point in the past 13.8 billion years. Not just lucky in a sentimental sense but a mathematical sense of several quintillion to one.

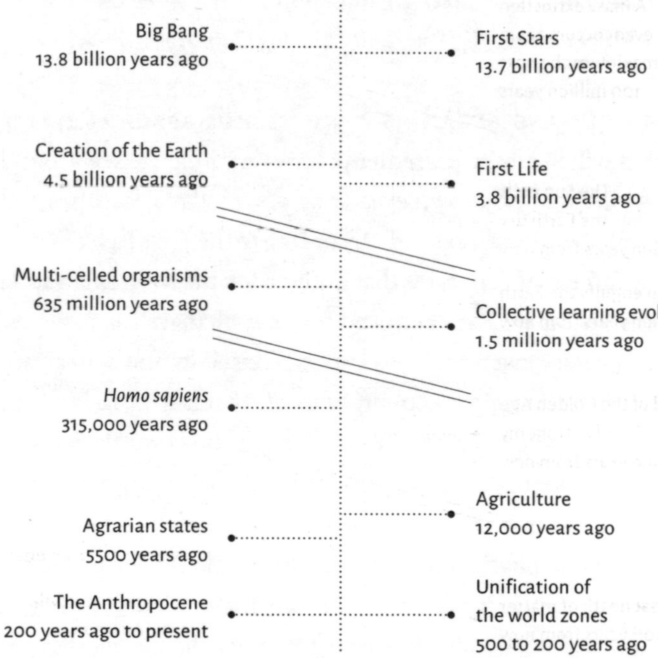
Big Bang
13.8 billion years ago
First Stars
13.7 billion years ago
Creation of the Earth
4.5 billion years ago
First Life
3.8 billion years ago
Multi-celled organisms
635 million years ago
Collective learning evolves
1.5 million years ago
Homo sapiens
315,000 years ago
Agriculture
12,000 years ago
Agrarian states
5500 years ago
Unification of
the world zones
500 to 200 years ago
The Anthropocene
200 years ago to present

The overarching trend of universal history is complexity, and the overarching trend of human history is collective learning – which in turn increases complexity. Using these two trends, we can make some forecasts about the future on short and long time scales. This is somewhat rare in the field of historical studies. Moreover, it is only in the future where the trends of complexity and collective learning really bear fruit and make the meaning of the upward trend clear for the reader.

So where exactly is our story likely to go next?

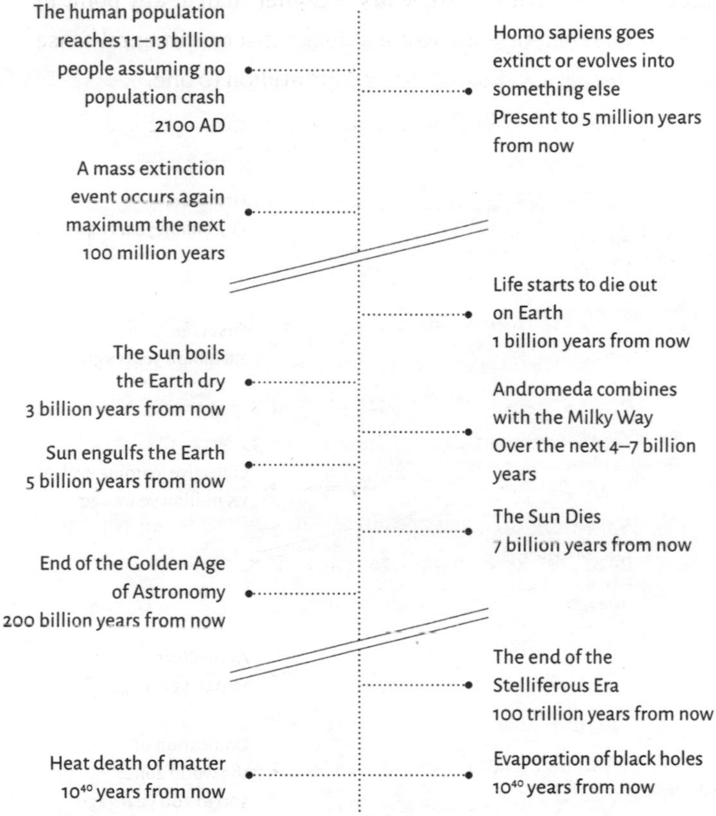
The human population reaches 11–13 billion people assuming no population crash
2100 AD
Homo sapiens goes extinct or evolves into something else
Present to 5 million years from now
A mass extinction event occurs again maximum the next 100 million years
Life starts to die out on Earth
1 billion years from now
The Sun boils the Earth dry
3 billion years from now
Andromeda combines with the Milky Way
Over the next 4–7 billion years
Sun engulfs the Earth
5 billion years from now
The Sun Dies
7 billion years from now
End of the Golden Age of Astronomy
200 billion years from now
The end of the Stelliferous Era
100 trillion years from now
Heat death of matter
10^{40} years from now
Evaporation of black holes
10^{40} years from now

PREDICTING THE FUTURE

When predicting the future, you must not predict one future but several. And then assess each scenario based on their plausibility. These multiple futures, regardless of their exact detail, fall on a spectrum.

1. **Projected future:** What science says is happening. Things play out how current trends suggest they play out. It is business as usual, where we assume no major change to variables or behaviours, and no dynamic discoveries. The projected future may not even be the most likely future, since new discoveries and changes in variables do eventually occur, but it forms an important baseline for our forecasting. For instance, a projected future would involve the outcome of greenhouse gas emissions and global industrial growth continuing at current rates.

2. **Probable future:** What science says could happen. Where variation or change within the bounds of known science indicates where trends might go. The probable future is the projected future's margin for error, or margin for variation. A probable future refers to something that science already understands but which hasn't yet come to pass: for example, transition to and heavier reliance on solar technologies and less reliance on fossil fuels.

3. **Possible future:** What science might discover. Where a discovery as yet unknown to science alters a future outcome, or where we can't scientifically

explain in detail how everything will work. We are not visionary engineers capable of predicting technological progress 200 years from now. Imagine how difficult it would have been in 1800 AD to predict the existence of the internet, or its societal effects. A possible future has an unknown variable like an algebra equation: 'present + x = outcome'. In fact, like an algebra equation, we can use the known variables to get a clearer picture of what the value of x actually is. Major advances in AI or nuclear fusion or quantum computing innovations (which we do not yet know how to fully engineer) would fall into this category.

4. **Preposterous future:** What science says can't happen. Where an outcome seems to openly defy the laws of known science, contradicting all available data or understanding. It plays an important role in prediction because it clearly defines what a possible future is by mere contrast. It prevents speculation about technology going too overboard. But it can also serve to predict technologies that are currently too mind-boggling. The Moon landing might have seemed preposterous to someone in 1800 AD, before rocketry or even human flight. An example of a preposterous future today is a technology that defies the Second Law of Thermodynamics.

In fact, on a long enough timescale, complexity can turn the preposterous into the possible, then the probable, and even the projected. If nothing else, the only way to figure out the limits of the possible is by going beyond them to the impossible.

ANALYSING THE NEAR FUTURE

It is actually easier to predict the Deep Future, which has timescales of billions and trillions of years, than it is to predict the Near Future on timescales of hundreds or thousands of years. This is entirely down to complexity. Cosmological changes in the wider Universe that take billions of years to occur deal with *relatively* simple systems and calculations. Provided we have the right data, we know how long the Sun will live and how long it will take Andromeda to combine with the Milky Way, despite the fact that those things are billions of years away. Humanity is a much more complex system. Each individual human is capable of billions of different behaviours. In aggregate of billions of people, this makes for quite the calculation – one which no supercomputer is capable of. It is very difficult to predict what inventions humanity will stumble upon, and how these inventions will affect people's behaviour in society. Finally, humanity's interactions with nature (something that is also very complex) make it difficult to predict the rise of diseases or natural disasters.

Even though the events of the next century are difficult to predict, all the possible outcomes of the Near Future over the next 100 to 300 years fall into four broad categories. They relate to whether human complexity rises, stabilises, gracefully decreases or collapses.

1. **Technological breakthrough**: Where human society does not hit a limit to its current modes of production in the next 100 to 300 years and rates of innovation keep pace with growth of the human population. Perhaps it involves the economically viable distribution of nuclear fusion power, making energy cheap enough even for

the poorest countries to develop, with an exponential increase on the limits of energy and production globally, and without the corresponding degradation of the biosphere that comes with fossil fuels. But such a breakthrough also includes those scenarios where humanity hands the reins of future complexity to artificial intelligence. That is, collective learning sets off another jump in complexity.

2. **Green equilibrium**: Where human society over the next 100 to 300 years does not develop a major technological breakthrough in the Near Future (by no means a guarantee, since the first agriculture and Industrial Revolution are 12,000 years apart) and lives within its means to avoid total degradation of the biosphere. This may include technological innovation at a smaller scale, along with some good planning, government policy and a shift to more sustainable forms of production. Human complexity does not increase significantly, nor does it decrease.

3. **Creative descent:** Where human society invokes a form of policy that actually reduces human production and consumption in order to ward off environmental or demographic disaster. It is a deliberate unravelling of human complexity. Examples of scenarios within this category include radical population control and reduction, dismantling of heavy industry, restrictions on car and air travel, restrictions on energy consumption and production rather than their replacement with renewable forms, rationing of food

and clothing, and so on. Over a long enough period of descent, human complexity more closely resembles the agrarian states of 300 years ago than society today.

4. **Collapse:** Involving every conceivable doomsday scenario. Environmental disaster, nuclear war, superbugs, an asteroid impact or a super-volcanic eruption. This category covers every scenario where human complexity dramatically declines, regardless of the cause.

Take a moment and ask yourself which future is the most likely? Why? For nearly two decades, the public discussions about climate change have prompted a notable rise in pessimism in developed countries. After two years of a global pandemic that resulted, among other things, in job losses and a spike in mental health issues, that pessimism may well have increased.

However, on a long enough timescale, collective learning will prompt another breakthrough in technology. Human complexity simply has to endure the scenario that occurs for long enough without collapsing for this to take effect. In that sense, humanity's great task in the twenty-first century is to survive it. In all likelihood, we shall, which means that complexity might continue to increase for many millennia to come – with all the astounding new breakthroughs that implies.

In the wider context of rising complexity in the Universe, what happens in the twenty-first century may determine whether the trend continues or ends here. In that sense, the generations alive today and those that will be born in the next few years stand at a pivotal moment in history. It is a period where human actions will have a magnified impact, much more than any king, peasant, farmer or forager spread across the last 315,000 years. In a very real material and temporal sense, what you do in your life matters, and potentially may echo into the future – in a way that very few individual actions have done before.

A hydrogen reactor

THE 'NATURAL' DEEP FUTURE

Analysis of the Deep Future falls into two broad streams. The first is the 'natural' projected/probable futures of the Earth and the Universe, where higher complexity like biology or society have no impact on the processes of cosmology. The second stream is a series of possible/preposterous futures where complexity continues to increase for millions, billions and even trillions of years beyond the current stage of human technology on Earth to the point where the wider cosmos is affected and manipulated by us.

According to current data, here are the natural projected/ probable things that will happen in the Deep Future:

1. **1 billion years from now, death of the biosphere:** Mass extinction events happen every 100 million years on average. But so far they have not yet succeeded in ending the world, just in wiping out a large percentage of existing species. The Deep Future is much more certain. In about a billion years, the Sun will begin to exhaust its fuel. Its luminosity will increase, CO_2 levels will decrease, and this means plants on Earth over the following years will find it harder and harder to do most forms of photosynthesis and thus sustain complex life on our tiny rock. Life would struggle and decline from the 1-billion-year mark onward. That's nearly twice the amount of time that separates us from the Cambrian Explosion 541 million years ago. That's a great deal of time for multi-celled species to continue to evolve and change for a period of time nearly *double* what separates us from our jawless fish vertebrate ancestors. Even if humans go extinct, it is entirely possible that another species capable of collective learning could evolve during that time and then in a few hundred thousand years match or surpass our current level of complexity.

2. **3 to 7 billion years from now, death of the Earth and the Sun:** At the 3-billion-year mark, the Sun will grow larger and larger until it boils the surface of the Earth dry. Once we get to an Earth's surface with a temperature greater than 100°C (212°F), we can be pretty sure that's it for life on Earth. Perhaps some

single-celled organisms could still exist in the cracks of the Earth, but that is a clear decline of complexity and the end of the tale in our biosphere. Then the Sun will grow so large it will engulf the Earth, burning and absorbing whatever is left. The planet itself will be destroyed. The Sun may also bloat up to destroy Mars. But it will never get so large that it goes beyond that, leaving the Asteroid Belt and the gas giants largely unscathed. After that, the Sun will shrink back and eventually extinguish itself. If our descendants are still around in such a massive number of years, we are likely to be incredibly advanced in technology to the point of being godlike. We will either have left the Earth to terraform and live on the moons of Jupiter and Saturn, or we may well have macro-engineered the Sun so it has a replenished supply of hydrogen to burn, or we may have left the solar system for other planets, abandoned the galaxy entirely or evolved to somehow not require a planet to live on at all.

3. **200 billion years from now, end of the Golden Age of Astronomy:** As dark energy continues to accelerate the expansion of the Universe past the speed of light, we would no longer get to see the light from other galaxies. If we were to lose the knowledge of Big Bang cosmology, our galaxy would be all we'd see. Or think exists. We'd revert to the idea that the Universe had no start date, is static and eternal. The Milky Way would be our entire Universe. That is why a number of scientists refer to the current age where we can see evidence for the Big Bang, and can see other galaxies, as the Golden

Age of Astronomy. We are lucky to have been born at such an early stage in the Universe, where, relatively speaking, the cosmos is still in its infancy, being only 13.8 billion years old in a Universe that will live on for many trillions of years.

4. **100 trillion years from now, the end of stars:** Once the Universe is trillions of years old, all new star formation will have ceased in all the galaxies of the Universe, and only the smallest slow-burning stars will still be burning. By the time we reach 100 trillion years from now, the last of these tiny, slow-burning stars will have died. At this point there will be no conventional energy flows to sustain life on planets, and it will become difficult for any moderately advanced seafaring societies to find enough energy flows to sustain or increase their complexity. One alternative would be to use the radiation that comes from black holes, but this does not emit in such generous amounts as a star. The saving grace of this projection is the astounding lengths complexity could have reached after trillions of years of collective learning (or whatever more rapid process supersedes it).

5. **10^{40} years from now, heat death of matter:** This is a 1 with forty zeros, or 10 duodecillion years from now. Or to put it another way, a trillion multiplied by a trillion, multiplied by *another* trillion, plus four more zeros. Or to put it yet another way, nearly three times the length of time that separates us from the end of stars. At this point, not only will stars have disappeared but the very fabric of planets and asteroids will have

crumbled. All molecular combinations in the Universe will have long since decayed and only singular atoms will remain. Except these will have gradually decayed into simpler and simpler atoms too. Once we are down to just hydrogen atoms, these too will decay back into energy and the Universe will be nothing but a void filled with weak radiation becoming more and more evenly distributed thanks to the Second Law of Thermodynamics. The energy flows that created complexity so far in our story will have finished their work and all complexity in the Universe will have come to an end. This is what I meant when I said that the Second Law is at once the creator and destroyer of worlds. What we will have left is a blank eternity, with no change, no events, no history. Not just the end of the world, but the end of our story. The end of all history. 10^{40} years from now, even black holes will have emitted all their radiation and evaporated into thinly distributed energy as well.

This scenario is known as the 'Big Freeze' and is the projected end to the history of complexity in the Universe according to our current data. This is built on the idea that the Universe will continue accelerating and expand forever.

There are also two probable futures for the end of the Universe, if we observe new rates of expansion in the Universe which alter the data we use to make predictions. If the Universe is accelerating faster in expansion than we currently observe, then we have the 'Big Rip' scenario, where the Universe expands so quickly it increases the space between galaxies, then overpowers the force of gravity and tears galaxies apart, then overpowers the nuclear forces that hold atoms together, and rips stars, planets and organisms apart. This could happen in as little as 20 billion years from now. I say 'little', but that is still a heck of a long time.

The second probable future is the 'Big Crunch', where the acceleration of the expansion of the Universe actually slows down and reverses itself, eventually squeezing all galaxies in the Universe back together, and then further back into the white-hot singularity with which our story began. If this sets off yet another Big Bang, this leads to the corollary 'Big Bounce' scenario, where the Universe expands again and is reborn over and over. Very poetic and appealing. Data doesn't currently reflect this, but if the expansion of the Universe slowed down and reversed itself, this could take somewhere between 50 and a few hundred billion years.

As grim as the 'Big Freeze' scenario may sound, with its 'dying with a whimper' aesthetic, it actually gives complexity the maximum amount of time (trillions upon trillions upon trillions of years) to continue to increase and find a solution to the mortality imbued in the Universe by the Second Law of Thermodynamics. In that sense, we should actually be popping the champagne that the 'Big Freeze' thus far seems to be the most likely outcome of our story.

THE DEEP FUTURE OF COMPLEXITY

Consider how 'young' the 13.8-billion-year Universe is in terms of the 100 trillion years it will exist before all the stars burn out, and the trillions upon trillions upon trillions of years before the heat death of matter. Consider the tiny amount of time that biological complexity has existed on Earth (3.8 billion years) and the even more infinitesimal amount of time that humanity has existed in literate states and societies (5500 years). Finally, consider how far collective learning and scientific progress has accelerated in the past 200 years.

That is such a small percentage of the amount of time that the Universe will exist that it is negligible. It is not even worth expressing in a percentage, given all the zeros that will go in front of it. And if complexity continues to accelerate (as it currently is), then once we start to consider where it could go in thousands or millions of years, to speak nothing of billions or trillions, then a highly advanced society could conceivably impact the natural evolution of the Universe.

Assuming that complexity continues to increase, this outcome is not only possible but it gradually becomes probable or even projected.

Yet it is almost impossible to predict what such advanced complexity would look like. Humans have difficulty speculating what technology will look like a decade from now, much less what it will look like on the timescale of thousands or millions of years. But there is one way in which we can gain an idea of just how complex and powerful those super-civilisations would be someday.

At the beginning of this book, we explored a metric for complexity – the density of energy flows that create, sustain and increase complexity. The Sun achieves 2 erg/g/s, an average photosynthesising organism 900 erg/g/s, a dog 20,000 erg/g/s, human foraging societies 40,000 erg/g/s, agrarian states 100,000 erg/g/s, nineteenth-century industrial society 500,000 erg/g/s and society today 2 million erg/g/s. With a quantifiable metric such as this, we are able to project how complex super-civilisations would be in the Deep Future, and even estimate how long it would take us to reach that point.

With each increase in energy flows came an increase in the structural intricacy of complexity: from a blob of hydrogen atoms to a single cell with DNA, to multi-celled organisms composed of a network of trillions of cells, to the networks of humans, their domestic animals and all our machines which form a society. Also, at each increase of energy density comes the ability of humans, at least, to consciously manipulate the laws of physics and alter the environment around them to ensure their continued survival.

Without having the faintest idea of what science or miraculous inventions would arise in such super-civilisations, the trends we have so far observed in our story imply that complexity could become so advanced that it would start influencing the structure of galaxies and cosmological evolution itself.

SUPER-CIVILISATIONS

In 1964, Russian astronomer Nikolai Kardashev proposed a scale by which we could evaluate advanced civilisations based on how much energy they harnessed. The various stages on this scale are measured by harnessing the equivalent of all the energy of a planet, star or galaxy. Note that does not necessarily mean that a planet, star or galaxy must be the actual source of such energy, but simply that a super-civilisation has somehow generated the *equivalent* of such energy.

1. **Type I Civilisation (Planet):** In the Anthropocene, humans are actually quite close to harnessing the energy equivalent of an entire planet. We are currently a '0.7' civilisation or higher. So if we project a little into the future, our average free energy rate density would be 2,600,000 erg/g/s. In fact, that is such a small

increase in average energy flows for society one can quite adequately speculate what such a society would look like: a planet without much larger a population, with more abundant energy sources to sustain their complexity (for example, a population of 10 billion with nuclear fusion reactors all living at a standard of living akin to the developed world today or better). If we project the accelerating rate of increasing complexity for humanity from the foraging era to today, the calculation indicates that humanity will achieve a Type I civilisation in 300 years or less. Purely by looking at the numbers, humanity's future looks fairly *bright*. Assuming we can keep complexity from backsliding. This is precisely why the current generations alive today are so pivotal in our story.

2. **Type II Civilisation (Star):** At this point, we have moved from projected and probable futures to the possible futures where science does not yet have the precise knowledge to explain what technology would get us to this stage. The stage at which humanity (or whatever we turn into) harnesses the energy equivalent of a star may conjure up images of a Dyson sphere. This is where we surround a star with panels that absorb all the energy it gives off, rather than the small fraction of the energy it throws off into the entire Universe and hits plants or solar panels or other energy sources here on Earth. The free energy density of a super-civilisation that harnesses the energy equivalent of a star is approximately 70,200,000,000 erg/g/ – a major leap forward in complexity compared to modern society.

And considerably more structurally intricate and capable of manipulating the surrounding environment and/or the fundamentals of physics in the Universe. That is roughly the same degree of difference in complexity that separates a single-cell organism from a World War II Spitfire engine. It is distinctly possible by this point that humans may be 'trans-human' or 'post-human'. Perhaps humans will have managed to reverse the effects of biological aging, or even uploaded their consciousnesses to computers to live on forever, either as a collective consciousness or individual cyborgs, with such advanced calculating power that collective learning, communication and new inventions blaze along at blinding speed. Again, according to the current rate of acceleration of complexity, this would take us another 25,000 years at most. 25,000 years ago, human foragers were spread across Africa, Europe, Asia and Australasia. It is roughly twice the space of time that separates us from agriculture. In terms of the total life expectancy of complexity in the Universe, even 25,000 years is an insignificant fraction of trillions upon trillions upon trillions of years. Even if we just consider the 100 trillion years where stars will still burn, that is 0.0000000025 per cent of the time we would have to reach this stage.

What these numbers make clear is something astrobiologists and SETI-enthusiasts (those engaged in the search for extra-terrestrial intelligence) already suspected: that it may take billions of years for complexity to get started in the Universe, but once it gets going, the time between each breakthrough gets smaller and smaller and smaller.

Dyson sphere

This leaves an insane amount of time for complexity to rise somewhere in the Universe – even if it doesn't happen with our particular species.

3. **Type III Civilisation (Galaxy):** If our hypothetical super-civilisation finds that harnessing the power of a single star isn't enough to manipulate the most fundamental laws of physics in the Universe, then they could always move on to harness the energy equivalent of the 200 to 400 billion stars that exist in the Milky Way galaxy. Such a powerful super-civilisation would have 14,000,000,000,000,000,000,000,000 erg/g/s (14 septillion) in free energy density. That is greater in terms of complexity than the difference between a single subatomic particle and modern society. Such a super-civilisation would essentially make our entire

society and its powers look as complex as a quark. At this point we are talking about a society that has indisputably godlike powers to the point that it could probably manipulate an entire galaxy, if not the fundamental laws of the Universe itself, to suit its interests. If we use the same rate of acceleration of complexity, as massive as this number is, this could be achieved in less than 100,000 years. The same amount of time that separates us from *Homo sapiens* first migrating out of Africa. Even if these projections are off, physicists have previously estimated it would take between 5 and 50 million years for us to reach every solar system in the galaxy (assuming it is impossible to move faster than the speed of light). Or roughly the same amount of time that separates us from our last common ancestor with chimpanzees or with primates. Even 50 million years is a tiny fragment of the amount of time life on Earth has existed, much less the length of time that stars and galaxies will continue to exist.

If we literally harness the stars in a galaxy to achieve such power levels, we may well have to move stars around into some sort of 'energy grid'. This is called galactic macro-engineering. If there is extremely advanced life elsewhere in the Universe that has already beat us to this point, we may be wrong in looking for radio signals from other societies. We may want to be looking among the 400 billion galaxies out there for signs of galactic structures that seem to have no natural explanation.

4. **Type IV Civilisation (Universe):** We are now firmly in the realm of the preposterous future. While it may be physically possible to travel across the Milky Way galaxy, we would need some physics-defying technology to enter every single galaxy in the Observable Universe. If somehow this was achieved, however, we would harness approximately 6,000,000,000,000,000,000,000,000, 000,000,000,000 erg/g/s (6 uno-dectillion). There is no point of comparison for complexity here. We exhausted that with Type III civilisations. Nothing exists in the Universe so far that is so simple or so complex that we could contrast this in complexity to our current society. But given we have a numerical value, we can still calculate how long it would take us to reach this level. The results of the calculation are surprising. Assuming we can overcome the many physical and technological barriers with our previous godlike Type II and Type III civilisations, we could achieve Type IV in 200,000 years or less. According to this calculation, we'd go from our current Type 0.7 to a Type IV civilisation in roughly 325,000 years. Just a little bit more than the period of time that *Homo sapiens* has been known to exist, and an extremely small fragment of time during which complexity can exist in the Universe. Even if these calculations are way off and the acceleration of complexity slows down significantly somewhere along the line, we can afford to be off by almost *nine orders of magnitude* before all the stars in the Universe burn out.

It is highly unlikely a super-civilisation would need to harness this much power in order to achieve environment-manipulating ability to bend or break the physical laws of the Universe. It is highly likely such an ability would already be achieved at Type II or Type III.

5. **Type V Civilisation (Multiverse):** Since we have come this far, we might as well go all the way. Assuming that a so-called Multiverse, as described in Chapter 1, exists, and that it is possible to somehow traverse an eternity of inflationary space and unite all the energy flows from universes (where there exists such a thing as energy) in some sort of network (which would require us to thoroughly bend the properties of space and time), a Type V civilisation harnesses all the energy flows of all the universes out there. Unfortunately, it is impossible to give a free energy number for this. Not just because the number would be insanely large. The number may well be infinite if the number of universes in the Multiverse is infinite. And without a finite number, it is impossible to project how long it would take, since traversing an infinite amount of universes would take an infinite amount of time. In that sense, if complexity could ever possibly get that far, we'd achieve what is essentially a 'singularity' of complexity, where it runs towards infinity, past an event horizon of invention, where anything and everything is possible.

If the statement that we probably wouldn't need to go this far to manipulate the fundamental properties of our Universe applies to Type IV civilisations, then that goes infinitely more times over for a Type V.

THE 'BIG SAVE'

We have so far covered the three outcomes for the 'natural' end of the Universe, where complexity has no impact on cosmological evolution: the Big Freeze, the Big Rip and the Big Crunch/Bounce. Presumably all advanced civilisations in these scenarios don't get much further than their own planets and eventually just go extinct. Which anyone will tell you is distinctly possible.

But the alternative futures where complexity continues to accelerate and does not suddenly stop at some point reveals one final endgame for our universal story. The scenario where a super-civilisation that is a Type II, Type III or Type IV actually grows so capable at manipulating its surrounding environment that it can somehow defy the Second Law of Thermodynamics and extend the life of complexity beyond its natural end date. In other words, the 'Big Save'.

Given how quickly such a super-civilisation could be reached, relative to the current age of the Universe or how much time the Universe has to exist in the Big Rip, Big Crunch/Bounce and especially the Big Freeze scenarios, it is not unreasonable to add this possibility to the list.

Consider that in the last 635 million years that multicellular species have existed on Earth, at least one species out of an estimated 10 billion of them has produced collective learning to such a degree that societies were created. And most of that work was done in the past 12,000 years. Many astrobiologists agree that there could be up to 300 million habitable planets in the Milky Way galaxy. Assuming they all produce multi-celled life (which they wouldn't), that still would mean that an organism capable of collective learning would be unlikely to arise somewhere else in the Milky Way. However, when you also factor in the number of galaxies in the Observable Universe (approximately 400 billion), if you assume an average of 300 million habitable planets per galaxy, then the odds of another species being capable of becoming a super-civilisation get quite a lot higher. And when you consider that there are trillions of years for such a species to arise again, when it took only 13.8 billion for us to arrive, the odds get overpowering. Even if humanity goes extinct somewhere in the Near Future (a possibility one can sense simply by turning on the news these days), there are good odds that a Type II, Type III or Type IV Civilisation could arise elsewhere in the Universe.

That is why the Big Save has to be considered in any horizon scan done for the endgame of the Universe, alongside the much more predictable natural endings.

In a Big Save scenario, we would attain either a Type II, III or IV super-civilisation (whatever was required technologically) and engage in one of three activities to prolong our complexity beyond the projected natural end of the Universe:

1. **Escape:** Assuming a Multiverse exists, we could simply leave for a universe that wasn't quite so old or whose physical properties did not include the Second Law of Thermodynamics killing complexity by exhausting all energy flows.

2. **Manipulate:** Assuming a Multiverse does not exist or it is physically impossible to travel to other 'coffee-cup rings' on the beige table of the cosmos, a powerfully complex super-civilisation might be able to manipulate the fundamental properties of the Universe (or rewrite them) in order to defy the Second Law of Thermodynamics. This

seems to be the most likely case. A form of technology that generates perpetual motion (locally or universally) in order to undo the natural end of things.

3. **Create:** Most compatible with the Big Crunch, but not exclusive of the Big Freeze or Big Rip, if we could somehow manipulate space-time, we'd simply recreate the Big Bang, but pre-coded with conditions that produce physical laws and a distribution of matter and energy much friendlier to complexity than our own Universe.

All of the Big Save scenarios fall into the realm of the 'preposterous' future, since they require not just an invention of something we don't currently understand (a 'possible' future) but achieving something that currently science says would be physically impossible. Yet it is only by riding the line of the preposterous that we find out what truly is possible.

If we look at the short amount of time it would take to achieve the level of a super-civilisation, the long life ahead of complexity in the Universe and the immense complexity of those super-civilisations as shown by the numbers, it is worth bearing in mind. Things that would have seemed preposterous to humans living only a few centuries ago – instantaneous communication, travelling faster than the speed of sound or landing on the Moon – have been achieved by modern society. The demand on us isn't that heavy: just to stick around for another 20,000 to 300,000 years and see what might happen.

THE ANSWER MAY NOT BE 42

I write at a time of extreme hardship and pessimism globally. Our population is experiencing strain (not entirely due to a pandemic but certainly worsened by it) and the worst political factionalism in living memory. A person of an anxious disposition may think this sounds suspiciously and dangerously like the downturn of a secular cycle.

It is therefore with great pleasure that I can talk not only about the Anthropocene but the end of the Universe with a great deal of optimism, buoyancy and hope. The same patterns that have propelled us through the history of all existence seem to imply good and honest chances of surviving not only in the Near Future but into the Deep Future as well. And not just surviving but thriving. Perhaps even unravelling more of the great mysteries of the Universe. That is the supreme potential of human society, knowledge and endeavour. It is extremely valuable.

Our actions in this day and age may hold the secret to launching a magnificent array of astounding things, which, on the timescales of the history of the Universe, are very close to hand indeed. And if any amount of optimism is to be applied to longevity technology or transhumanism, it is possible that either we or our children will be able to take part in that great adventure firsthand. That is the tremendous gift that all the exertions of the past have handed to us and that we may pass on to others.

We have explored at a leap and a bound the history of 13.8 billion years. But this story may only just be beginning.

Be brave, be good to each other.

Acknowledgements

I would like to thank David Christian for training me, giving me innumerable opportunities and sticking by me in good and bad times, particularly in the most current disasters brought on by the pandemic.

I would also like to thank my parents, Susan and Greg Baker, for their boundless support and patience, and for supporting me in pursuit of a rather unusual field.

I'd like to thank Jason Gallate for the crucial moral support these past several months and for reading through drafts of this book. You literally saved my life.

In that respect I'd also like to thank Karen Stapley and Matt Diteljan for reading through drafts of this book and giving useful feedback that vastly improved its quality.

Lastly, I'd like to thank Milo. He knows why.

Further Reading

Adas, Michael. *Islamic and European Expansion: The Forging of a Global Order*. Philadelphia: Temple University Press, 2001.

Adshead, S., *China in World History*. 2nd edn. Basingstoke: Macmillan, 1995.

Allen, Robert. *The British Industrial Revolution in a Global Perspective*. Cambridge: Cambridge University Press, 2009.

Allsen, Thomas. *Culture and Conquest in Mongol Eurasia*. Cambridge: Cambridge University Press, 2001.

Alvarez, Walter. *A Most Improbable Journey: A Big History of Our Planet and Ourselves*. New York: W.W. Norton, 2016.

Alvarez, Walter. *T. Rex and the Crater of Doom*. Princeton: Princeton University Press, 1997.

Archer, Christon, et al. *World History of Warfare*. Lincoln: University of Nebraska Press, 2002.

Ashton, T.S. *The Industrial Revolution, 1760–1830*. London: Oxford University Press, 1948.

Asimov, Isaac. *Beginnings: The Story of Origins – of Mankind, Life, the Earth, the Universe*. New York: Walker, 1987.

Bairoch, Paul. *Cities and Economic Development: From the Dawn of History to the Present*. Trans. Christopher Brauder. Chicago: University of Chicago Press, 1988.

Baker, David. 'Collective learning: A potential unifying theme of human history'. *Journal of World History*, vol. 26, no. 1, 2015, pp. 77–104.

Barfield, Thomas. The Nomadic Alternative. Englewood Cliffs: Prentice-Hall, 1993.

Barnett, S.A. *The Science of Life: From Cells to Survival*. Sydney: Allen & Unwin, 1998.

Barrow, John. *The Book of Universes: Exploring the Limits of the Cosmos*. London: W.W. Norton, 2011.

Bayley, Chris. *The Birth of the Modern World: Global Connections and Comparisons, 1780–1914*. Oxford: Blackwell, 2003.

Bellwood, Peter. *First Famers: The Origins of Agricultural Societies*. Oxford: Blackwell, 2005.

Bentley, Jerry. *Old World Encounters: Cross-Cultural Contacts and Exchanges in Pre-Modern Times*. Oxford: Oxford University Press, 1993.

Berg, Maxine. *The Age of Manufacturers, 1700–1820: Industry, Innovation, and Work in Britain*. 2nd edn. London: Routledge, 1994.

Bin Wong, Robert. *China Transformed: Historical Change and the Limits of European Experience*. Ithaca: Cornell University Press, 1997.

Biraben, J.R. 'Essai sur l'évolution du nombre des homes'. *Population*, vol. 34, 1979, pp. 13–25.

Black, Jeremy. *War and the World: Military Power and the Fate of Continents, 1450–2000*. New Haven: Yale University Press, 1998.

Blackwell, Richard J. *Behind the Scenes at Galileo's Trial*. Notre Dame: University of Notre Dame Press, 2006.

Bowler, Peter. *Evolution: The History of an Idea*. 3rd edn. Berkeley: University of California Press, 2003.

Brantingham, P.J. et al. *The Early Paleolithic Beyond Western Europe*. Berkeley: University of California Press, 2004.

Bray, Francesca. *The Rice Economies: Technology and Development in Asian Societies*. Oxford: Basil Blackwell, 1986.

Brown, Cynthia. *Big History: From the Big Bang to the Present*. New York and London: The New Press, 2007.

Browne, Janet. *Charles Darwin: Voyaging*. Princeton: Princeton University Press, 1996.

Bryson, Bill. *A Short History of Nearly Everything*. New York: Broadway Books, 2003.

Bucciantini, Massimo, Michele Camerota and Franco Gudice. *Galileo's Telescope: A European Story*. Trans. Catherine Bolton. Cambridge, Mass.: Harvard University Press, 2015.

Cavalli-Sforza, Luigi Luca, and Francesco Cavalli-Sforza. *The Great Human Diasporas*. Trans. Sarah Thorne. Reading: Addison-Wesley, 1995.

Chaisson, Eric. *Epic of Evolution: Seven Ages of the Cosmos*. New York: Columbia University Press, 2006.

Chaisson, Eric J. *Cosmic Evolution: The Rise of Complexity in Nature*. Cambridge: Harvard University Press, 2001.

Chaisson, Eric. 'Using complexity science to search for unity in the natural sciences'. In Charles Lineweaver, Paul Davies and Michael Ruse (eds). *Complexity and the Arrow of Time*. Cambridge: Cambridge University Press, 2013.

Chambers, John and Jacqueline Morton. *From Dust to Life: The Origin and Evolution of Our Solar System*. Princeton: Princeton University Press, 2014.

Cheney, Dorothy and Robert Seyfarth. *Baboon Metaphysics: The Evolution of a Social Mind*. Chicago: University of Chicago Press, 2014.

Chi, Z. and H.C. Hung. 'The emergence of agriculture in South China'. *Antiquity*, vol. 84, 2010, pp. 11–25.

Christian, David. 'The evolutionary epic and the chronometric revolution'. In Genet et al. (eds) *The Evolutionary Epic: Science's Story and Humanity's Response*. Santa Margarita: Collingswood Foundation Press, 2009.

Christian, David. *Maps of Time: An Introduction to Big History*. Berkeley: University of California Press, 2004.

Christian, David. *Origin Story: A Big History of Everything*. London: Allen Lane, 2018.

Christian, David. 'Silk Roads or Steppe Roads? The Silk Roads in World History'. *Journal of World History*, vol. 11., no. 1 (2000), pp. 1–26.

Christian, David and Cynthia Stokes Brown and Craig Benjamin. *Big History: Between Nothing and Everything*. New York: McGraw Hill, 2014.

Christianson, Gale. *Edwin Hubble: Mariner of the Nebulae*. Chicago: University of Chicago Press, 1996.

Cipolla, Carlo. *Before the Industrial Revolution: European Society and Economy, 1000–1700*. 2nd edn. London: Methuen, 1981.

Cloud, Preston. *Oasis in Space: Earth History from the Beginning*. New York: W.W. Norton, 1988.

Coe, Michael. *Mexico: From the Olmecs to the Aztecs*. 4th edn. New York: Thames and Hudson, 1994.

Cohen, Mark. *Health and the Rise of Civilization*. New Haven: Yale University Press, 1989.

Collins, Francis. *The Language of Life: DNA and the Revolution in Personalised Medicine*. London: Profile Books, 2010.

Copernicus, Nicolaus. 'De hypothesibus motuum coelestium a se constitutis commentariolus' in *Three Copernican Treatises*. 2nd edn. Trans. Edward Rosen. New York: Dover Publications, 2004.

Copernicus, Nicolaus. *De revolutionibus orbium coelestium*. Ed. trans. Edward Rosen. Baltimore: Johns Hopkins University Press, 1992.

Cowan, C. and P. Watson, eds. *The Origins of Agriculture: An International Perspective*. Washington: Smithsonian Institution Press, 1992.

Crawford, Harriet. *Sumer and the Sumerians*. Cambridge: Cambridge University Press, 2004.

Crosby, Alfred. *The Columbian Exchange: The Biological Expansion of Europe, 900–1900*. Cambridge: Cambridge University Press, 1986.

D'Altroy, Terence. *The Incas*. Malden: Blackwell, 2002.

Darwin, Charles. *The Autobiography of Charles Darwin 1809–1882*. Ed. Nora Barlow. London: Collins, 1958.

Darwin, Charles. *The Origin of Species by Means of Natural Selection*. 1st edn, reprint. Cambridge, Mass: Harvard University Press, 2003.

Darwin, Charles. *The Voyage of the Beagle*. New York: Cosimo Classics, 2008.

Davies, Kevin. *Cracking the Genome: Inside the Race to Unlock DNA*. Baltimore: Johns Hopkins University Press, 2001.

De Waal, Frans. *Chimpanzee Politics: Power and Sex Among Apes*. Johns Hopkins University Press, 2007.

De Waal, Frans. *Tree of Origin: What Primate Behaviour Can Tell Us about Human Social Evolution*. Cambridge: Harvard University Press, 2001.

Diamond, Jared. *Guns, Germs, and Steel: The Fates of Human Societies*. London: Vintage, 1998.

Dunbar, Robin. *A New History of Mankind's Evolution*. London: Faber & Faber, 2004.

Dunn, Ross. *The Adventures of Ibn Battuta: A Muslim Traveler of the Fourteenth Century*. Berkeley: University of California Press, 1986.

Dyson, Freeman. *Origins of Life*. 2nd edn. Cambridge: Cambridge University Press, 1999.

Earle, Timothy. *How Chiefs Come to Power: The Political Economy in Prehistory*. Stanford: Stanford University Press, 1997.

Ehret, Christopher. *An African Classical Age: Eastern and Southern Africa in World History, 1000 BC to AD 400*. Charlottesville: University Press of Virginia, 1998.

Ellis, Walter. *Ptolemy of Egypt*. London: Routledge, 1994.

Elvin, Mark. *The Pattern of the Chinese Past.* Stanford, Calif.: Stanford University Press, 1973.

Erwin, Douglas. *Extinction: How Life on Earth Nearly Ended 250 Million Years Ago*. Princeton: Princeton University Press, 2006.

Fagan, Brian. *People of the Earth: An Introduction to World Prehistory.* 10th edn. New Jersey: Prentice Hall, 2001.

Faser, Evan and Andrew Rimas. *Empires of Food: Feast, Famine, and the Rise and Fall of Civilisations.* Berkeley, Calif.: Counterpoint, 2010.

Fernandez-Armesto, Felipe. *Before Columbus: Exploration and Colonisation from the Mediterranean to the Atlantic, 1229–1492.* London: Macmillan, 1987.

Fernandez-Armesto, Felipe. *Pathfinders: A Global History of Exploration*. New York: W.W. Norton, 2007.

Flannery, Tim. *The Future Eaters: An Ecological History of the Australasian Lands and People.* Chatswood: Reed, 1995.

Fortey, R. *Earth: An Intimate History.* New York: Knopf, 2004.

Frankel, Henry. *The Continental Drift Controversy: Wegener and the Early Debate.* Cambridge: Cambridge University Press, 2012.

Galilei, Galileo. *Dialogue Concerning Two Chief World Systems: Ptolemaic and Copernician*. Trans. Stillman Drake. Ed. Stephen Jay Gould. Berkeley: University of California Press, 2001.

Gates, Charles. *Ancient Cities: The Archaeology of Urban Life in the Ancient Near East*, Egypt, Greece, and Rome, 2nd edn. Abingdon: Routledge, 2011.

Ghorsio, A. et al. 'New elements einsteinium and fermium, atomic numbers 99 and 100'. *Physical Review*, vol. 99, no. 3 (1955), pp. 1048–9.

Gingerich, Owen. *Copernicus: A Very Short Introduction*. Oxford: Oxford University Press, 2016.

Goodall, Jane. *The Chimpanzees of Gombe: Patterns of Behaviour.* Cambridge: Harvard University Press, 1986.

Goodall, Jane. *Through a Window: My Thirty Years with the Chimpanzees of Gombe*. Boston: Houghton Mifflin, 1990.

Gordin, Michael. *A Well Ordered Thing: Dmitrii Mendeleev and the Shadow of the Periodic Table*. New York: Basic Books, 2004.

Gosling, Raymond (interview). 'Due credit'. *Nature*, vol. 496 (2013). Available from www.nature.com/news/due-credit-1.12806

Green, R. et al. 'A draft sequence of the neanderthal genome'. *Science*, vol. 328, no. 5979 (May 2010), pp. 710–22.

Hansen, Valerie. *The Open Empire: A History of China to 1600*. New York: W.W. Norton, 2000.

Hawking, Stephen. *A Brief History of Time: From the Big Bang to Black Holes*. New York: Bantam, 1988.

Hawking, Stephen and Leonard Mlodinow. *The Grand Design*. New York: Bantam Books, 2010.

Hawking, Stephen. *The Universe in a Nutshell*. New York: Bantam, 2001.

Hazen, Robert. *The Story of Earth: The First 4.5 Billion Years from Stardust to Living Planet*. New York: Viking 2012.

Headrick, Daniel. *The Tools of Empire: Technology and European Imperialism in the Nineteenth Century*. New York: Oxford University Press, 1981.

Headrick, Daniel. *Technology: A World History*. Oxford: Oxford University Press, 2009.

Heilbron, John. *Galileo*. Oxford: Oxford University Press, 2010.

Higman, B. *How Food Made History*. Chichester: Wiley Blackwell, 2012.

Hoskin, Michael. *Discoverers of the Universe: William and Caroline Herschel*. Princeton: Princeton University Press, 2011.

Hsu, Cho-yun. *Han Agriculture: The Formation of Early Chinese Agrarian Economy, 206 B.C.–220 A.D.* Ed. Jack Dulled. Seattle: University of Washington Press, 1980.

Hubble, Edwin. 'A relation between distance and radial velocity among extra-galactic nebulae'. *Proceedings of the National Academy of Sciences*, vol. 15, no. 3 (1929), pp. 168–73.

Johanson, Donald, and Maitland Edey. *Lucy: The Beginnings of Humankind*. New York: Simon & Schuster, 1981.

Johnson, A. and T. Earle. *The Evolution of Human Societies: From Foraging Group to Agrarian State*. 2nd edn. Stanford: Stanford University Press, 2000.

Johnson, George. *Miss Leavitt's Stars: The Untold Story of the Woman Who Discovered How to Measure the Universe*. New York: W.W. Norton, 2005.

Jones, Rhys. 'Fire stick farming'. *Australian Natural History* (Sep. 1969), pp. 224–8.

Jordanova, Ludmilla. *Lamarck*. Oxford: Oxford University Press, 1984.

Karol, Paul et al. 'Discovery of the element with atomic number Z=118 completing the 7th row of the periodic table (IUPAC Technical Report)'. *Pure Applied Chemistry*, vol. 88 (2016), pp. 155–60.

Kenyon, Kathleen. *Digging up Jericho.* London: Ernest Benn, 1957.

Kicza, John. 'The peoples and civilizations of the Americas before contact'. *Agricultural and Pastoral Societies in Ancient and Classical History*. Ed. Michael Adas. Philadelphia: Temple University Press, 2001.

King, Henry. *The History of the Telescope*. New York: Dover Publications, 2003.

Klein, Richard. *The Dawn of Human Culture*. New York: Wiley, 2002.

Knoll, Andrew. *Life on a Young Planet: The First Three Billion Years of Evolution on Earth*. Princeton: Princeton University Press, 2003.

Korotayev, A., A. Malkov and D. Khalturina. *Laws of History: Mathematical Modelling of Historical Macroprocesses*. Moscow: Komkniga, 2005.

Krauss, Lawrence. *A Universe from Nothing: Why There is Something Rather than Nothing*. New York: Simon & Schuster, 2012.

Lamarck, Jean Baptiste Pierre Antoine de Monet. *Philosophie zoologique: ou exposition des considerations relatives a l'histoire naturelle des animaux*. Cambridge: Cambridge University Press, 2011.

Leakey, R. *The Sixth Extinction: Patterns of Life and the Future of Humankind*. New York: Doubleday, 1995.

Leavitt, Henrietta S. '1777 Variables in the Magellanic Clouds'. *Annals of Harvard College Observatory*, vol. 60, no. 4 (1908), pp. 87–108.

Leick, Gwendolyn. *Mesopotamia: The Invention of the City*. London: Penguin, 2001.

Levathes, Louise. *When China Ruled the Seas: The Treasure Fleet of the Dragon Throne, 1405–1433*. New York: Simon & Schuster, 1994.

Livi-Bacci, Massimo. *A Concise History of World Population*. Trans. Carl Ipsen. Oxford: Blackwell, 1992.

Lunine, J. *Earth: Evolution of a Habitable World.* Cambridge: Cambridge University Press, 1999.

Macdougall, Doug. *Why Geology Matters: Decoding the Past, Anticipating the Future.* Berkeley: University of California Press, 2011.

Maddison, Angus. *The World Economy: A Millennial Perspective*. Paris: OECD, 2001.

Maddox, Brenda. 'The double helix and the "wronged heroine"'. *Nature*, vol. 421 (2003), pp. 407–8.

Marcus, Joyce. *Mesoamerican Writing Systems: Propaganda, Myth, and History in Four Ancient Civilizations*. Princeton: Princeton University Press, 1992.

Marks, Robert. *The Origins of the Modern World: A Global and Ecological Narrative from the Fifteenth to the Twenty-First Century*. 2nd edn. Lanham: Rowman & Littlefield, 2007.

Maynard Smith, John and Eors Szathmary. *The Origins of Life: From the Birth of Life to the Origins of Language*. Oxford: Oxford University Press, 1999.

McBrearty, Sally and Alison Brooks. 'The revolution that wasn't: A new interpretation of the origin of modern human behaviour'. *Journal of Human Evolution*, 39 (2000), pp. 453–63.

McGowan, Christopher. *The Dragon Seekers: How an Extraordinary Circle of Fossilists Discovered the Dinosaurs and Paved the Way for Darwin*. London: Basic Books, 2009.

McNeill, J.R. and William H. McNeill. *The Human Web: A Bird's-Eye View of World History*. New York: W.W. Norton, 2003.

McNeill, William. *Plagues and People*. Oxford: Blackwell, 1977.

Mendeleev, Dmitri. 'Remarks concerning the discovery of gallium'. In *Mendeleev on the Periodic Law: Selected Writings, 1869–1905*. Ed. William Jensen. New York: Dover Publications, 2005.

Newton, Isaac. *The Mathematical Principles of Natural Philosophy*. Trans. Andrew Motte. London: Benjamin Motte, 1729.

Nicastro, Nicholas. *Circumference: Eratosthenes and the Ancient Quest to Measure the Globe*. New York: St Martin's Press, 2008.

Nutman, Allen et al. 'Rapid emergence of life shown by discovery of 3,700-million-year-old microbial structures'. *Nature*, vol. 537 (Sep. 2016), pp. 535–8.

Otfinoski, Steven. *Marco Polo: to China and Back*. New York: Benchmark Books, 2003.

Overton, Mark. *Agricultural Revolution in England: The Transformation of the Agrarian Economy, 1500–1850*. Cambridge: Cambridge University Press, 1996.

Pacey, Arnold. *Technology in World Civilisation*. Cambridge, Mass.: MIT Press, 1990.

Parker, Geoffrey. *The Military Revolution: Military Innovation and the Rise of the West, 1500–1800*. 2nd edn. Cambridge: Cambridge University Press, 1996.

Pinker, Steven. *The Blank State: The Modern Denial of Human Nature*. New York: Penguin, 2003.

Polo, Marco. *The Travels of Marco Polo*. Trans. Aldo Ricci. Reprint. Abingdon: Routledge Curzon, 2005.

Pomeranz, Kenneth. *The Great Divergence: China, Europe, and the Making of the Modern World Economy*. Princeton: Princeton University Press, 2000.

Pomeranz, Kenneth and Steven Topik. *The World that Trade Created: Society, Culture, and the World Economy, 1400 to the Present*. 2nd edn. Armonk: Sharpe, 2006.

Ponting, Clive. *A Green History of the World: The Environment and the Collapse of Great Civilisations*. London: Penguin, 1991.

Ptolemy, Claudius. *Ptolemy's Almagest*. Trans. and ed. G. Toomer. Princeton: Princeton University Press, 1998.

Ptolemy, Claudius. *Ptolemy's Geography: An Annotated Translation of the Theoretical Chapters*. Trans. and eds J. Berggren and Alexander Jones. Princeton: Princeton University Press, 2000.

Rampino, Michael and Stanley Ambrose. 'Volcanic winter in the garden of eden: the toba super-eruption and the Late Pleistocene population crash'. In *Volcanic Hazards and Disasters in Human Antiquity*. Ed. F. McCoy and W. Heiken. Boulder, Colo.: Geological Society of America, 2000, pg. 78–80.

Richards, John. *The Unending Frontier: Environmental History of the Early Modern World*. Berkeley: University of California Press, 2006.

Ringrose, David. *Expansion and Global Interaction, 1200–1700*. New York: Longman, 2001.

Ristvet, Lauren. *In the Beginning: World History from Human Evolution to the First States*. New York: McGraw-Hill, 2007.

Roller, Duane. *Ancient Geography: The Discovery of the World in Classical Greece and Rome*. London: I.B. Tauris, 2015.

Rothman, Mitchell. *Uruk, Mesopotamia, and Its Neighbours: Cross-Cultural Interactions in the Era of State Formation*. Santa Fe: School of American Research Press, 2001.

Rudwick, Martin. *Earth's Deep History: How It Was Discovered and Why It Matters*. Chicago: University of Chicago Press, 2014.

Russell, Peter. *Prince Henry the Navigator: A Life.* New Haven: Yale University Press, 2000.

Sahlins, Marshall. 'The original affluent society' In *Stone Age Economics.* London: Tavistock, 1972, pp. 1–39.

Sayre, A. *Rosalind Franklin and DNA.* New York: W.W. Norton, 1975.

Scarre, Chris, ed. *The Human Past: World Prehistory and the Development of Human Societies.* London: Thames & Hudson, 2005.

Schamandt-Besserat, Denise. *How Writing Came About: Handbook to Life in Ancient Mesopotamia.* Austin: University of Texas Press, 1996.

Sharratt, Michael. *Galileo: Decisive Innovator.* Cambridge: Cambridge University Press, 1994.

Smil, Vaclav. *Energy in World History.* Boulder: Westview Press, 1994.

Smith, Bruce. *The Emergence of Agriculture.* New York: Scientific American Library, 1995.

Strayer, Robert. *Ways of the World: A Global History.* Boston: St Martin's Press, 2009.

Stringer, Chris. *The Origin of Our Species.* London: Allen Lane, 2011.

Tarbuck, E. and F. Lutgens. *Earth: An Introduction to Physical Geology.* New Jersey: Pearson Prentice Hall, 2005.

Tattersall, Ian. *Masters of the Planet: The Search for Human Origins.* New York: Palgrave Macmillan, 2012.

Tattersall, Ian. *Becoming Human: Evolution and Human Uniqueness.* New York: Harcourt Brace, 1998.

Temple, Robert. *The Genius of China: 3000 Years of Science, Discovery, and Invention.* New York: Touchstone, 1986.

Turchin, Peter and Sergei Nefedov. *Secular Cycles.* Princeton: Princeton University Press, 2009.

Venter, J. Craig. *A Life Decoded: My Genome, My Life.* London: Penguin, 2007.

Watson, Fred. *Stargazer: The Life and History of the Telescope.* Cambridge, Mass.: Da Capo Press, 2006.

Watson, James. *The Double Helix: A Personal Account of the Discovery of the Structure of DNA.* London: Atheneum Press, 1968.

Wegener, Alfred. *The Origin of Continents and Oceans.* Trans. John Biram. New York: Dover Publications, 1966.

Weinberg, Steven. *The First Three Minutes: A Modern View of the Origin of the Universe.* New York: Basic Books, 1977.

Westfall, Richard. *The Life of Isaac Newton*. Cambridge: Cambridge University Press, 1993.

Wilkins, Maurice. *The Third Man of the Double Helix: An Autobiography*. Oxford: Oxford University Press, 2005.

Woods, Michael and Mary Woods. *Ancient Technology: Ancient Agriculture from Foraging to Farming*. Minneapolis: Runestone Press, 2000.

Wrangham, Richard. 'The evolution of sexuality in chimpanzees and bonobos'. *Human Nature*, vol. 4 (1993), pp. 47–79.

Wrangham, Richard and Dale Peterson. *Demonic Males: Apes and the Origins of Human Violence.* Boston: Mariner Books, 1996.

Wrigley, E. *Energy and the English Industrial Revolution*. Cambridge: Cambridge University Press, 2011.

Zheng, Y. et al., 'Rice fields and modes of rice cultivation between 5000 and 2500 BC in East China'. *Journal of Archaeological Science*, vol. 36 (2009), pp. 2609–16.

Image Credits

Aira Pimping: Illustrations on pages 14, 22, 33, 35, 67, 126, 128, 146

Alan Laver: Illustrations on pages 110, 115, 118, 144, 162, 185

p. 13 NASA: WMAP Science Team / Science Photo Library

p. 24 NASA/JPL-Caltech / R. Hurt (SSC/Caltech) via Wikimedia Commons

p. 37 Mikkel Juul Jensen / Science Photo Library

p. 44 Mark Garlick / Science Photo Library

p. 45 Gary Hincks / Science Photo Library

p. 55 CNX Openstax via Wikimedia commons

p. 72 Nicolas Primola / Shutterstock

p. 73 Sebastian Kaulitzki / Alamy

p. 75 Liliya Butenko / Shutterstock

p. 77 Gwen Shockey / Science Photo Library

p. 78 Friedrich Saurer / Science Photo Library

p. 79 Nobumichi Tamura, Stocktrek Images / Alamy

p. 81 Richard Bizley / Science Photo Library

p. 82 Michael Long / Science Photo Library

p. 84 tinkivinki / Shutterstock.

p. 87 Sebastian Kaultzki / Science Photo Library

p. 90 Universal Images Group North America LLC / Alamy

p. 98 S. Entressangle / E. Daynes / Science Photo Library

p. 101 DK Images / Science Photo Library

p. 134 Rebecca Rose Flores / Alamy

p.141 Asian and Middle Eastern Division / New York Public Library / Science Photo Library

p. 166 Paul Fürst, Copper engraving of Doctor Schnabel [i.e Dr. Beak], a plague doctor in seventeenth-century Rome, with a satirical macaronic poem, c.1656, via Wikimedia Commons

p. 185 Interfoto / Alamy

p. 199 Science Photo Library / Alamy

p. 209 Kevin Gill via Wikimedia Commons

Index